More Science From an Easy Chair

RAY LANKESTER

Published in 1920

TABLE OF CONTENTS

PREFACE
A DAY IN THE OBERLAND
SWITZERLAND IN EARLY SUMMER
GLETSCH
THE PROBLEM OF THE GALLOPING HORSE
THE JEWEL IN THE TOAD'S HEAD
ELEPHANTS
A STRANGE EXTINCT BEAST
VEGETARIANS AND THEIR TEETH
FOOD AND COOKERY
SMELLS AND PERFUMES
KISSES
LAUGHTER
FATHERLESS FROGS
PRIMITIVE BELIEFS ABOUT FATHERLESS
PROGENY
THE PYGMY RACES OF MEN
PREHISTORIC PETTICOATS
NEW YEAR'S DAY AND THE CALENDAR
EASTERTIDE, SHAMROCKS AND SPERMACETI
MUSEUMS
THE SECRET OF A TERRIBLE DISEASE
CARRIERS OF DISEASE
IMMUNITY AND CURATIVE INOCULATIONS
THE STRANGE STORY OF ANIMAL LIFE IN NEW
ZEALAND
THE EFFACEMENT OF NATURE BY MAN
THE EXTINCTION OF THE BISON AND OF
WHALES
MORE ABOUT WHALES
MISCONCEPTIONS ABOUT SCIENCE

RAY LANKESTER

PREFACE

The present volume is a reprint of that issued in 1912 with the title, "Science from an Easy Chair: Second Series." It consists, like its predecessors, of chapters originally published by me in the Daily Telegraph, which I have revised and illustrated by a large number of drawings. In order to render the issue of the present cheap edition possible, it has been found necessary to restrict its size a little by the omission of chapters dealing with Glaciers, Ferns and Fern-seed, and the history of the Sea-squirts or Ascidians, which are contained in the original larger book. My hope is that this collection of papers, "about a number of things," may meet with as kind a reception from my readers as that which they have accorded to its predecessors.

E. RAY LANKESTER
July 1, 1920
MORE SCIENCE FROM AN EASY CHAIR

A DAY IN THE OBERLAND

I am writing in early September from Interlaken, one of the loveliest spots in Europe when blessed with a full blaze of sunlight and only a few high-floating clouds, but absolutely detestable in dull, rainy weather, losing its beauty as the fairy scenes of a theatre do when viewed by dreary daylight. It is the case of the little girl of whom it is recorded that "When she was good she was very good, and when she was not she was horrid." This morning, after four days' misconduct, Interlaken was very good. The tremendous sun-blaze seemed to fill the valleys with a pale blue luminous vapour, cut sharply by the shadows of steep hill-sides. Here and there the smoke of some burning weeds showed up as brightest blue. Far away through the gap formed in the long range of nearer mountains, where the Lütschine Valley opens into the vale of Interlaken, the Jungfrau appeared in full majesty, absolutely brilliant and unearthly. So I walked towards her up the valley. Zweilütschinen is the name given to the spot where the valley divides into two, that to the left leading up to Grindelwald, under the shadow of the Mönch and the Wetterhorn, that to the right bringing one to Lauterbrünnen and the Staubbach waterfall, with the snow-fields of the Tchingel finally closing the way—over which I climbed years ago to Ried in the Loetschen Thal.

The autumn crocus was already up in many of the closely trimmed little meadows, whilst the sweet scent of the late hay-crop spread from the newly cut herbage of others.

At Zweilütschinen, where the white glacier-torrent unites with the black, and the milky stream is nearly as cold as ice, and is boiling along over huge rocks, its banks bordered with pine forest, I came upon a native fishing for trout. He was using a short rod and a weighted line with a small "grub" as bait. He dropped his line into the water close to the steep bank, where some projecting rock or half-sunk boulder staved off the violence of the stream.

7

He had already caught half-a-dozen beautiful, red-spotted fish, which he carried in a wooden tank full of water, with a close-fitting lid to prevent their jumping out. I saw him take a seventh. The largest must have weighed nearly two pounds. It seems almost incredible that fish should inhabit water so cold, so opaque, and so torrential, and should find there any kind of nourishment. They make their way up by keeping close to the bank, and are able, even in that milky current, to perceive and snatch the unfortunate worm or grub which has been washed into the flood and is being hurried along at headlong speed. Only the trout has the courage, strength, and love of nearly freezing water necessary for such a life—no other fish ventures into such conditions. Trout are actually caught in some mountain pools at a height of 8,000 ft., edged by perpetual snow.

You are rarely given trout to eat here in the hotels. A lake fish, called "ferras," a large species of the salmonid genus Coregonus, to which the skelly, powan, and vendayce of British lakes belong, is the commonest fish of the table d'hôte, and not very good. A better one is the perch-pike or zander. It is common in all the larger shallow lakes of Central Europe, and abounds in the "broads" which extend from Potsdam to Hamburg, though it is unknown in the British Isles. It is quite the best of the European fresh-water fish for the table, and there should be no difficulty about introducing it into the Norfolk Broads. It would be worth an effort on the part of the Board of Agriculture and Fisheries to do so, as the perch-pike, unlike other fresh-water fishes, would hold its own on the market against haddock, brill, and plaice. Another interesting fresh-water fish which grows to a large size in the Lake of Geneva (where I have seen it netted) is the burbot—called "lote" in French—a true cod of fresh-water habit which, though common throughout Europe and Northern Asia, is, in our country, only taken in a few rivers opening on the east coast. It is a brilliantly coloured fish, orange-brown, mottled with black, and is very good eating.

Passing up the Lauterbrünnen valley, I came upon some wild raspberries and quantities of the fine, large-flowered sage, Salvia glutinosa, with its yellow flowers, in shape like those of the dead-nettle, but much bigger. They were being visited by humble-bees, and I was able to see the effective mechanism at work by which the bee's body is dusted with the pollen of the flower. I have illustrated this in some drawings (Fig. 1) which are accompanied by a detailed explanation. Two long stamens, a1, arch high up over the lip of the flower, li, on which the bee alights, and are protected by a keel or hood of the corolla. Each stamen is provided with a broad process, a2, standing out low down on its arched stalk, and blocking the way to the nectar in the cup of the flower. When the bee pushes his head against these obstacles and forces them backwards, the result is to swing the long arched stalk, with its pollen sacks, in the opposite direction, namely, forwards and downwards on to the bee's back. It was easy to see this movement going on, and the consequent dusting of the bee's back with pollen. In somewhat older flowers, which have been relieved of their pollen, the style, st., or free stalk-like extremity of the egg-holding capsule, already as long as the stamens, grows longer and bends down towards the lip or landing-place of the yellow flower. When a pollen-dusted bee alights on one of these maturer flowers the sticky end of the now depending style is gently rubbed by the bee's back and smeared with a few pollen-grains brought by the bee from a distant flower. These rapidly expand into "pollen tubes," or filaments, and, penetrating the long style, reach the egg-germs below. Thus cross-fertilization is brought about by the bees which come for the nectar of Salvia. The stalks and outer parts of the flower of this plant produce a very sticky secretion which effectually prevents any small insects from crawling up and helping themselves to the nectar exclusively provided for the attraction of the humble-bee, whose services are indispensable.

As I walked on, a belated Apollo butterfly, with its two red spots, and a pale Swallow-tail fluttered by me. Then some children emerged from unsuspected lurking-places in the wood and offered bunches of edelweiss (Fig. 2). This curious-looking little plant does not grow (as pretended by reporters of mountaineering disasters) exclusively in places only to be reached by a dangerous climb. I have gathered it in meadows on the hillside above Zermatt, and it is common enough in accessible spots. The flowers are like those of our English groundsel and yellow in colour—little "composite" knobs, each built up of many tubular "florets" packed side by side. Six or seven of these little short-stalked knobs of florets are arranged in a circlet around a somewhat larger knob, and each of them gives off from its stalk one long and two shorter white, hairy, leaf-like growths, flat and blade-like in shape and spreading outwards from the circle, so that the whole series resemble the rays of a star (or more truly of a star-fish!). They look strangely artificial, as though cut out of new white flannel (with a greenish tint), and have been dignified by the comparison of the shape of the white-flannel rays with that of the foot of the lion and the claws of the eagle. They are extraordinary-looking little plants, and are similar in their hairiness and pale tint to some of the seaside plants on our own coast, which, in fact, include species closely allied to them ("cud-weeds" of the genus Gnaphalium).

The huge cliffs of rocks on either side (in some parts over a thousand feet in sheer height from the torrent) come closer to one another in the part where we now are than in most Alpine valleys, so as almost to give it the character of a "gorge." At some points the highest part of the precipice actually overhangs the perpendicular face by many feet. A refreshing cold air comes up from the icy torrent, whilst the heat of the sun diffuses the delicious resinous scent of the pine trees. Above the naked rock we see steep hill-sides covered with forest, and away above these again bare grass-slopes topped by cloud. But as the clouds slowly lift and break we become suddenly aware of something impending far above and beyond all this, something more dazzling in its white brightness than the sun-lit clouds, a form sharply cut in outline and firm, yet rounded by a shadow of an exquisite purple tint which no cloud can assume. The steely blue Alpine sky fits around this marvel of pure whiteness as it towers through the opening cloud, and soars out of earth's range. What is this glory so remote yet impending over us? It is the Jungfrau, the incomparable virgin of the ice-world, who bares her snowy breast. She slowly parts her filmy veil, and, as we gaze, uncovers all her loveliness.

The rock walls of the Lauterbrünnen valley show at one place a thickness of many hundred feet of strongly marked, perfectly horizontal "strata"—the layers deposited immense ages ago at the bottom of a deep sea. Not only have they been raised to this position, and then cut into, so as to make the profound furrow or valley in the sides of which we see them, but they have been bent and contorted in places to an extent which is, at first sight, incredible. Close to one great precipice of orderly horizontal layers you see the whole series suddenly turned up at right angles, and the same strata which were horizontal have become perpendicular. But that is not the limit, for the upturned strata are seen actually to turn right over, and again become horizontal in a reversed order, the strata which were the lowest becoming highest, and the highest lowest. The rock is rolled up just as a flat disc of Genoese pastry—consisting of alternate layers of jam and sponge-cake—is folded on itself to form a double thickness. The forces at work capable of treating the solid rocks, the foundations of the great mountains, in this way are gigantic beyond measurement. This folding of the earth's crust is caused by the fact that the "crust," or skin of the earth, has ceased to cool, being warmed by the sun, and therefore does not shrink, whilst the great white-hot mass within (in comparison with which the twenty-mile-thick crust is a mere film) continually loses heat, and shrinks definitely in volume as its temperature sinks. The crust or jacket of stratified rock deposited by the action of the waters on the surface of the globe has been compelled—at whatever cost, so to speak—to fit itself to the diminishing "core" on which it lies. Slowly, but steadily, this "settlement" has gone on, and is going on. The horizontal rock layers, being now too great in length and breadth, adjust themselves by "buckling"—just as a too large, ill-fitting dress does—and the Alps, the Himalayas, and other great mountain ranges, are regions where this "buckling" process has for countless ages proceeded, slowly but surely. Probably the "buckling" has proceeded to a large extent without sudden movement, but with a lateral pressure of such power as ultimately to throw a crust of thousands of feet thickness into deep folds a mile or so in vertical measurement from crest to hollow, protruding from the general level both upwards and downwards, whilst often the folds are rolled over on to each other.

This crumbling and folding has gone on at great depths—that is to say, some miles below the surface (a mere nothing compared with the 8,000 miles diameter of the globe itself), though we now see the results exposed, like the pastry folded by a cook. Immense time has been taken in the process. A folding movement involving a vertical rise of an inch in ten years would not be noticed by human onlookers, but in 600,000 years this would give you a vertical displacement of more than 5,000 ft. (nearly a mile!). It has been shown that in Switzerland, along a line of country extending from Basle to Milan, strata of 10,000 ft. to 20,000 ft. in thickness, which, if straightened out, would give a flat area of that thickness, and of 200 miles in length, have been buckled and folded so as to occupy only a length of 130 miles! The former tight-fitting skin of horizontal rock layers has "had to" buckle to that extent here (and in the same way in other mountain ranges in other parts of the world), because the whole terrestrial sphere has shrunk, owing to the gradual cooling of the mass, whilst the crust has not shrunk, not having lost heat.

Filled with interest and delight in these things, I reached the railway station at Lauterbrünnen, from whence the little train is driven far up the mountain, even into the very heart of the Jungfrau, by an electric current generated by a turbine, itself driven by the torrent at our feet, the waters of which have descended from the glaciers far above, to which it will carry us. In a few minutes I was gently gliding in the train up the to the "Wengern Alp" and the "Little Scheidegg"—a slope up which I have so often in former years painfully struggled on foot for four hours or more. One could to-day watch the whole scene, in ease and comfort, during the two hours' ascent of the train. And a marvellous scene it is as one rises to the height of 8,000 ft., skirting the glaciers which ooze down the rocky sides of the Jungfrau, and mounting far above some of them. At the Scheidegg I changed into a smaller train, and with some thirty fellow-passengers was carried higher and higher by the faithful, untiring electric current. After a quarter of an hour's progress we paused high above the "snout" of the great Eiger glacier, and descended by a short path on to it, examined the ice, its crevasses and layers, and its "glacier-grains," and watched and heard an avalanche. The last time I was here it took a couple of hours to reach this spot from the Scheidegg, and probably neither I nor any of my fellow-passengers could to-day endure the necessary fatigue of reaching this spot on foot. Then we remounted the train, and on we went into the solid rock of the huge Eiger. The train stops in the rock tunnel and we got out to look, through an opening cut in its side, down the sheer wall of the mountain on to the grassy meadows thousands of feet below.

Then we start again, and on we are driven by the current generated away down there in Lauterbrünnen, through the spiral tunnel, mounting a thousand feet more till we are landed at an opening cut on the further side of the rocky Eiger, which admits us to an actual footing on the great glacier called the Eismeer, or Icelake. We lunch at a restaurant cut out as a cavern in the solid rock, and survey the wondrous scene. We are now at a height of 10,000 feet, and in the real frozen ice-world, hitherto accessible only to the young and vigorous. I have been there in my day with pain, danger, and labour, accompanied by guides and held up by ropes, but never till now with perfect ease and tranquillity and without "turning a hair," or causing either man or beast to labour painfully on my behalf. We had taken two hours only from Lauterbrünnen; in former days we should have started in the small hours of the morning from the Scheidegg, and have climbed through many dangers for some six or seven hours before reaching this spot.

I confess that I am not enchanted with all of the modern appliances for saving time and labour—the telegraph, the telephone, the automobile, and the aeroplane. But these mountain railways fill me with satisfaction and gratitude. When the Jungfrau railway was first projected, some athletic Englishmen with heavy boots and ice-axes, protested against the "desecration" of regions till then accessible only to them and to me, and others of our age and strength. They declared that the scenery would be injured by the railway and its troops of "tourists." As well might they protest against the desecration caused by the crawling of fifty house-flies on the dome of St. Paul's. These mountains and glaciers are so vast, and men with their railroads so small, that the latter are negligible in the presence of the former. No disfiguring effect whatever is produced by these mountain railways; the trains have even ceased to emit smoke since they were worked by electricity. I quite agree with those who object to "funiculars." The carriages on these are hauled up long, straight gashes in the mountain side, which have a hideous and disfiguring appearance. But I look forward with pleasure to the completion of the Jungfrau railway to the summit. I hope that the Swiss engineers will carry it through the mountain, and down along the side of the great Aletsch glacier to the Bel Alp and so to Brieg. That would be a glorious route to the Simplon tunnel and Italy!

I took three hours in the unwearied train descending from the Eismeer to Interlaken, and was back in my hotel in comfortable time for dinner, "mightily content with the day's journey," as Mr. Pepys would have said. I have always been sensitive to the action of diminished pressure, which produces what is called "mountain sickness" in many people. Many years ago I climbed by the glacier-pass known as the Weissthor from Macugnaga to the Riffel Alp, with a stylographic pen in my pocket. The reservoir of the pen contained a little air, which expanded as the atmospheric pressure diminished, and at 10,000 feet I found most of the ink emptied into my pocket. Probably one cause of the discomfort called "mountain sickness" arises from a similar expansion of gas contained in the digestive canal, and in the cavities connected with the ear and nose. The more suddenly the change of pressure is effected, the more noticeable is the discomfort. But I was rather pleased than otherwise to note, as I sat in the comfortable railway carriage, that when we passed 8,000 feet in elevation the old familiar giddiness, and tendency to sigh and gasp, came upon me as of yore, as I gathered was the experience of some of my fellow-passengers: and when we were returning, and had descended half-way to Lauterbrünnen, I enjoyed the sense of restored ease in breathing which I well remember when the whole experience was complicated by the fatigue of a long climb. A white-haired American lady was in the train with me ascending to the Eismeer. "I have longed all my life," she said, "to see a glaysher—to touch it and walk on it—and now I am going to do it at last. I and my daughter here have come right away from America to go on these cars to the glaysher." When we were descending, I asked the old lady if she had been pleased. "I can hardly speak of it rightly," she said. "It seems to me as though I have been standing up there on God's own throne." I do not sympathise with the Alpine monopolist who would grudge that dear old lady, and others like her, the little train and tramway by which alone such people can penetrate to those soul-stirring scenes. They are at least as sensitive to the beauty of the mountains as are the most muscular, most long-winded, and most sun-blistered of our friends—the acrobats of the rope and axe.

Interlaken

September, 1909

SWITZERLAND IN EARLY SUMMER

It is the early summer of 1910 and I have but just returned from a visit to Switzerland. The latter part of June and the beginning of July is the best for a stay in that splendid and happy land if one is a naturalist, and cares for the beauty of Alpine meadows, and of the flowers which grow among and upon the rocks near the great glaciers. This year the weather has, no doubt, been exceptionally cold and wet, and at no great height (5,000 feet) we have had snow-storms, even in July. But as compared with that of Paris and London the weather has been delightful. There has been an abundance of magnificent sunshine, and many days of full summer heat and cloudless sky. A fortnight ago (July 16th), and on the day before, it was as hot and brilliant in the valley of Chamonix as it can be. Mont Blanc and the Dome de Goutet stood out clear and immaculate against a purple-blue sky, and, as of old, we watched through the hotel telescope a party struggling, over the snow to the highest peak.

At Chillon the lake of Geneva, day after day, spread out to us its limitless surface of changing colour, now blending in one pearly expanse with the sky—so that the distant felucca boats seemed to float between heaven and earth—now streaked with emerald and amethystine bands. The huge mountain masses rising with a vast sweep from St. Jingo's shore displayed range after range of bloom-like greys and purples, whilst far away and above delicately glittered—like some incredible vision of a heavenly world beyond the sun-lit sky itself—the apparition of the snows and rocks of the great Dents du Midi. All this I have left behind me, and have passed back again to dull grey Paris, to the stormy Channel, and to the winter of London's July.

The incomparable pleasure which the lakes and valleys and mountains of Switzerland are capable of giving is due to the combination of many distinct sources of delight, each in itself of exceptional character. A month ago, in bright sunshine, I went, once again, by the little electric railway (most blessed invention of our day) from the pine-shaded torrent below to the great Eiger rock-mountain, and through its heart to the glacier beyond, more than 10,000 feet above sea-level. On the way back I left the train at the foot of the Eiger glacier, and walked down with my companion amongst the rocks of the moraine and over the sparse turf of these highest regions of life. Everywhere was a profusion of gentians, the larger and darker, as well as the smaller, bluest of all blue flowers. The large, plump, yellow globe-flowers (Trollius), the sulphur-yellow anemone, the glacial white-and-pink buttercup, the Alpine dryad, the Alpine forget-me-nots and pink primroses, the summer crocus, delicate hare-bells, and many other flowers of goodly size were abundant. The grass of Parnassus and the edelweiss were not yet in flower, but lower down the slopes the Alpine rhododendron was showing its crimson bunches of blossom. It is a pity that the Swiss call this plant "Alpenrose," since there is a true and exquisite Alpine rose (which we often found) with deep red flowers, dark-coloured foliage, and a rich, sweet-briar perfume. Lovely as these larger flowers of the higher Alps are, they are excelled in fascination by the delicate blue flowers of the Soldanellas, like little fringed foolscaps, by the brilliant little red and purple Alpine snap-dragon, and by the cushion-forming growths of saxifrages and other minute plants which encrust the rocks and bear, closely set in their compact, green, velvet-like foliage, tiny flowers as brilliant as gems. A ruby-red one amongst these is "the stalkless bladder-wort" (Silene acaulis), having no more resemblance at first sight to the somewhat ramshackle bladder-wort of our fields than a fairy has to a fishwife. There are many others of these cushion-forming, diminutive plants, with white, blue, yellow, and pink florets. Examined with a good pocket lens, they reveal unexpected beauties of detail—so graceful and harmonious that one wonders that no one has made carefully coloured pictures of them of ten times the size of nature, and published them for all the world to enjoy. Busily moving within their charmed circles we see, with our lens, minute insects which, attracted by the honey, are carrying the pollen of one flower to another, and effecting for these little pollen flowers what bees and moths do for the larger species.

Thus we are reminded that all this loveliness, this exquisite beauty, is the work of natural selection—the result of the survival of favourable variations in the struggle for existence. These minute symmetrical forms, this wax-like texture, these marvellous rows of coloured, enamel-like encrustation, have been selected from almost endless and limitless possible variations, and have been accumulated and maintained there as they are in all their beauty, by survival of the fittest—by natural selection. All beauty of living things, it seems, is due to Nature's selection, and not only all beauty of colour and form, but that beauty of behaviour and excellence of inner quality which we call "goodness." The fittest, that which has survived and will survive in the struggle of organic growth, is (we see it in these flowers) in man's estimation the beautiful. Is it possible to doubt that just as we approve and delightedly revel in the beauty created by "natural selection," so we give our admiration and reverence, without question, to "goodness," which also is the creation of Nature's great unfolding? Goodness (shall we say virtue and high quality?) is, like beauty, the inevitable product of the struggle of living things, and is Nature's favourite no less than man's desire. When we know the ways of Nature, we shall discover the source and meaning of beauty, whether of body or of mind.

As these thoughts are drifting through our enchanted dream we suddenly hear a deep and threatening roar from the mountain-side. We look up and see an avalanche falling down the rocks of the Jungfrau. The vast mountain, with its dazzling vestment of eternal snow, and its slowly creeping, green-fissured glaciers, towers above into the cloudless sky. In an instant the mind travels from the microscopic details of organic beauty, which but a moment ago held it entranced, to the contemplation of the gigantic and elemental force whose tremendous work is even now going on close to where we stand. The contrast, the range from the minute to the gigantic, is prodigious yet exhilarating, and strangely grateful. How many millions of years did it take to form those rocks (many of them are stratified, water-laid deposits) in the depths of the ocean? How many more to twist and bend them and raise them to their present height? And what inconceivably long persistence of the wear and tear of frost and snow and torrent has it required to excavate in their hard bosoms these deep, broad valleys thousands of feet below us, and to leave these strangely moulded mountain peaks still high above us? And that beauty of the sun-lit sky and of the billowy ice-field and of the colours of the lake below and of the luminous haze and the deep blue shade in the valley—how is that related to the beauty of the flowers? Truly enough, it is not a beauty called forth by natural selection. It is primordial; it is the beauty of great light itself. The response to its charm is felt by every living thing, even by the smallest green plant and the invisible animalcule, as it is by man himself. As I stand on the mountain-side we are all, from animalcule to man, sympathizing and uniting, as members of one great race, in our adoration of the sun. And in doing this we men are for the moment close to and in happy fellowship with our beautiful, though speechless, relatives who also live. Even the destructive bacteria which are killed by the sun probably enjoy an exquisite shudder in the process which more than compensates them for their extinction.

The pleasures of flower-seeking in Switzerland are by no means confined to the great heights. At moderate heights (4,000 to 5,000 feet) you have the Alpine meadows, and below those the rich-soiled woods which fill in the sides of the torrent-worn valleys. You cannot see an Alpine meadow after July, as it is cut down by then. It is at its best in June. It bears very little grass, and consists almost entirely of flowers. In places the hare-bells and Canterbury bells and the bugloss are so abundant as to make a whole valley-floor blue as in MacWhirter's picture. But more often the blue is intermixed with the balls of, red clover and the spikes of a splendid pale pink polygonum (a sort of buckwheat) and of a very large and handsome plantain. Large yellow gentians, mulleins, the nearly black and the purple orchids, vetches of all colours, the Alpine clover with four or five enormous flowers in a head instead of fifty little ones, the Astrantias (like a circular brooch made up of fifty gems each mounted on a long elastic wire and set vibrating side by side), the sky-blue forget-me-nots, and the golden potentillas, are usually components of the Alpine meadow. At Murren, and no doubt commonly elsewhere, there are a few very beautiful grasses among the flowers, but the most remarkable grass is one (Poa alpina), which has on every spikelet or head a bright green serpent-like streamer. Each of these "streamers" is, in fact, a young grass-plant, budded off "viviparously," as it is called, from the flower-head, or "spikelet," and having nothing to do with the proper fertilized seed or grain. The young plants so budded fall to the ground, and striking root rapidly, grow into separate individuals. It is probably owing to some condition in Alpine meadows adverse to the production of fertilized seed that this viviparous method of reproduction has been favoured, since it occurs also in an Alpine meadow-plant allied to the buckwheat, namely, Polygonum viviparum (not the kind mentioned above), where the lower flowers are converted into little red bulbs, by which the plant propagates. Both the viviparous grass and the polygonum are found in England. In fact, a very large proportion of Alpine plants occur in parts of the British islands (a legacy from the glacial period), though many which are abundant in Switzerland are rare and local here.

At a lower level, in the woods, we come upon other plants, not really "Alpine" at all, but of great and special beauty. We found four kinds of winter-green (Pirola), one with a very large, solitary flower, white and wax-like, and the beautiful white butterfly-orchid with nectaries three quarters of an inch long, and other large-flowered orchids. We were anxious to find the noble Martagon lily, and hunted in many glades and forest borders for it. At last, concealed on a bank in a wood, between Glion and Les Avants, it revealed itself in quantity, many specimens standing over three feet in height. Martagon is an Arabic word, signifying a Turkish cap. A very strange and uncanny-looking lily, which I had never seen before, turned up near Kandersteg at the Blue Lake, beloved of Mr. H. G. Wells. This is "the Herb Paris." It has four narrow outstretched green sepals, and four still narrower green petals, eight large stamens, and a purple seed capsule. Its broad oval leaves are also arranged in whorls of four. Its name has nothing to do with the "ville lumière," nor with the Trojan judge of female beauty, but refers to the symmetry and "parity" of its component parts. I was not surprised to find that "the Herb Paris" is poisonous, and was anciently used in medicine. It looks weird and deadly.

Marmots, glacier fleas (spring-tails, not true fleas), admirable trout, and burbot (the fresh-water cod, called "lote" in French), outrageous wood-gnats, which English people call by a Portuguese name as soon as they are on the Continent, and singing birds (usually one is too late in the season to hear them) were our zoological accompaniment. There were singularly few butterflies or other insects, probably in consequence of the previous wet weather.

July, 1909

GLETSCH

Varied and uncertain as the weather was in Switzerland during July of the year 1910, it showed a more decided character when I returned there at the end of August. For three weeks there was no flood of sunshine, no blazing of a cloudless blue sky, which is the one condition necessary to the perfection of the beauty of Swiss mountains, valleys and lakes. The Oberland was grey and shapeless, the Lauterbrünnen valley chilly and threatening; even the divine Jungfrau herself, when not altogether obliterated by the monotonous, impenetrable cloud, loomed in steely coldness—"a sterile promontory." Crossing the mountains from the Lake of Thun, we came to Montreux, only to find the pearl-like surface of the great Lake Leman transformed into lead. Not once in eight days did the celestial fortress called Les Dents du Midi reveal its existence, although we knew it was there, immensely high and remote, far away above the great buttresses of the Rhone valley. So completely was it blotted out by the conversion of that most excellent canopy, the air, into a foul and pestilent congregation of vapours, that it was difficult to imagine that it was still existing, and perhaps even glowing in sunshine above the pall of cloud. Italy, surely, we thought, would be free from this dreadful gloom.

The southern slopes of the Alps are often cloudless when the colder northern valleys are overhung with impenetrable mist. In four hours you can pass now from the Lake of Geneva through the hot Simplon Tunnel to the Lago Maggiore. So, hungering for sunshine, we packed, and ran in the ever-ready train through to Baveno. Thirty years ago we should have had to drive over the Simplon—a beautiful drive, it is true—but we should have taken sixteen hours in actually travelling from Montreux, and have had to pass a night en route at Brieg! A treacherous gleam of sunshine lasting half an hour welcomed us on emerging from the Simplon tunnel, and then for eight days the same leaden aspect of sky, mountain, and lake as that which we had left in Switzerland was maintained. Even this could not spoil altogether the beauty and interest of the fine old garden of the Borromeo family on the Isola Bella. Really big cypress trees, magnificent specimens of the Weymouth pine—the white pine of the United States, Pinus strobus, first brought from the St. Lawrence in 1705, and planted in Wiltshire by Lord Weymouth—a splendid camphor tree, strange varieties of the hydrangea, and many other old-fashioned shrubs adorn the quaint and well-designed terraces of that seat of ancient peace. The granite quarries close behind Baveno, and the cutting and chiselling of the granite by a population of some 2,000 quarrymen and stonemasons, were not deprived of their human interest by rain and skies more grey than the granite itself. But, at last, we gave up Italy in despair, retreated through the tunnel one morning, and an hour after mid-day were careering in a carriage along the Rhone valley—with jingling of bells and much cracking of a harmless whip—upwards on a drive of seven hours to the Rhone glacier, to the hotel called "Gletsch," staking all on the last chance of a change in the weather.

We passed the enclosed meadow near Brieg, whence three days later the splendidly daring South-American aviator started on his flight across the Alps, only to die after victory—a hero, whose courage and fatal triumph were worthy of a better cause. After some hours, passing many a black-timbered mountain village—the houses of which, set on stone piles, are the direct descendants of the pile-supported lake dwellings of the Stone Age on the shores of the Lake of Neuchatel—we came to the upper and narrower part of the valley. The road ascended by zig-zags through pine forests, in which the large blue gentian, with flowers and leaves in double rows on a gracefully bowed stem, were abundant. In open places the barberry, with its dense clusters of crimson fruit, was so abundant as actually to colour the landscape, whilst a huge yellow mullen nearly as big as a hollyhock, and bright Alpine "pinks," were there in profusion. Before the night fell, a long, furry animal, twice the size of a squirrel, and of dark brown colour, crossed the road with a characteristic undulating movement, a few feet in front of our carriage. It was a pine-marten, the largest of the weasel and pole-cat tribe, still to be found in our own north country. It must not be confused with the paler beech-marten of Anne of Brittany, which often takes up its abode in the roofs of Breton houses, according to my own experience in Dinard and the neighbourhood. Night fell, and our horses were still toiling up the mountain road. Impenetrable chasms lay below, and vast precipices above us. We crossed a bridge, and seemed in the darkness to plunge into the sheer rock itself, and, though thrilled with a delightful sense of mystery and awe, were feeling a little anxiety at the prospect of another hour among these gloomy, intangible dangers, when we rounded a projecting rock, and suddenly a brilliant constellation burst into view in the sky. It was the electric outfit of the Belvedere Hotel, 7,500 feet above the sea, and far up more than a thousand feet above us and the glacier's snout. In another minute the great arc lamps of the Gletsch Hotel, close to us, blazed forth, and we were welcomed into its snug hall and warmed by the great log-fire burning on its hospitable hearth.

The next day we were early afoot in the most brilliant sunshine, under a cloudless sky—really perfect Alpine weather. In the shade the persisting night-frost told of the great height of the marvellous amphitheatre which lay before us. The valley by which we had mounted the previous night abruptly abandons its steep gradient and gorge-like character, and widens into a flat, boulder-strewn plain, a little over a mile in diameter, surrounded, except for the narrow gap by which we had entered, by the steep, rocky sides of huge mountains. At the far end of the plain, a mile off, the great Rhone glacier comes toppling over the precipice, a snowy white, frozen cascade of a thousand feet in height. It looks even nearer than it is, and the gigantic teeth of white ice at the top of the fall seem no bigger than sentry-boxes, though we know they are more nearly the size of church steeples. The celebrated Furca road zig-zags up the mountain side for a thousand feet close to the glacier, and when you drive up it and reach the height of the Belvedere, you can step on to the ice close to the road. Then you can mount on to the flat, unbroken surface of the broad glacier stream above the fall, and trace the glacier to the snow-covered mountain-tops in which it originates. There is no such close and intimate view of a glacier to be had elsewhere in Europe by the traveller in diligence or carriage. We walked by the side of the infant Rhone, among the pebbles and boulders, to the overhanging snout of the great glacier from beneath which the river emerges. A very beautiful wine-red species of dwarf willow-herb (Epilobium Fleischeri) was growing abundantly in tufts among the pebbles, and many other Alpine plants greeted our eyes. The heat of the sun was that of midsummer, whilst a delicate air of icy freshness diffused itself from the great frozen mass in front of us.

Some large blocks of the glacier ice had fallen from above, and lay conveniently for examination. Whilst the walls of the ice-caves which have been cut into this and other glaciers present a perfectly smooth, continuous surface of clear ice, these fragments which had fallen from the surface exposed to the heat of the sun, were, as seen in the mass, white and opaque. When a stick was thrust into the mass, it broke into many-sided lumps of the size of a tennis-ball, which separated, and fell apart in a heap, like assorted coals thrown from a scuttle, though white instead of black. These were the curious glacier nodules, "grains du glacier," or "Gletcherkörne," characteristic of glacier ice as contrasted with lake ice. This structure of the glacier ice is peculiar to it, and is only made evident where the sun's rays penetrate it and melt the less pure ice which holds together the crystalline nodules. According to Dr. J. Young Buchanan, these nodules are masses of ice crystals comparatively free from mineral matter, whilst the water around them, which freezes less readily, contains mineral impurities in solution. The presence of saline matter in solution lowers, in proportion to its amount, the freezing-point of the water. Accordingly, although frozen into one solid mass with the nodules, the cementing ice melts under the heat of the penetrating rays of the sun sooner—that is, at a lower temperature— than do the purer crystalline nodules, and allows them to separate. It is owing to this that the exposed surface of glacier ice is white and powdery, disintegrated by the superficial heat, and forming a rough surface, on which one can safely walk. Lake ice does not break up in this manner under the sun's rays, but as it melts retains its smooth, slippery surface. It is formed in water, and not from the cementing and regelation of the powdery crystalline snow, as is glacier ice.

Pictures of the Rhone glacier published in the year 1820 and in the eighteenth century show that in old days the terminal ice-fall did not end abruptly in a narrowed "snout," as it does now, but spread out into a very broad half-dome or fan-shaped, apron-like expanse, some 700 feet high and a quarter of a mile broad at the base. It was considered one of the wonders of Switzerland, and was pictured in an exaggerated way in travellers' books. In 1873, when I first drove down the Furka road and saw the Rhone glacier, this wonderful, apron-like, terminal expansion of the glacier was still in existence. It has now completely disappeared. In those days, and for many years later, there was only a mule-path over the adjacent Grimsel Pass, but now there is a carriage road leading out of the Rhone glacier's basin northwards to Meiringen, whilst the old-established Furka road, at the other side of the amphitheatre, leads eastward to Andermatt, the St. Gothard, and the Lake of Lucerne. Hence three great roads now meet at Gletsch. Before leaving this wondrous spot we inspected some plump marmots, who were leading a happy life of ease and plenty in a large cage erected in front of the hotel; then in absolutely perfect weather we mounted the Grimsel road. We heard the frequent whistling of uncaged marmots as we ascended, and saw many of the little beasts sitting up on the rocks and diving into concealing crevices as we approached, just as do their smaller but closely allied cousins the prairie marmots (so-called "prairie dogs") of North America. The view, as one ascends the Grimsel, of the snow-peaks around Gletsch is a fine one in itself, but is vastly enhanced in beauty by the plunge downwards of the rocky gorge made by the Rhone as it leaves the flat-bottomed amphitheatre of its birth. The top of the Grimsel Pass, which is a little over 7,000 feet above sea-level, is the most desolate and bare of all such mountain passes. The rock is dark grey, almost black, and of unusually hard character. It is unstratified, and so resistant that it is everywhere worn into smooth, rounded surfaces, instead of being splintered and shattered. A small, black-

looking lake at the top of the pass contains to this day the bones of 500 Austrians and French who fought here in 1799. It is called the Totensee, or Dead Men's Lake. At this point one stands on a great watershed, dividing the rivers of the north from the rivers of the south. You may put one foot in a rivulet which is carrying water down the Aar Valley, and through the Lakes of Brienz and of Thun to the Rhine and North Sea, whilst you keep the other in another little stream, whose particles will pass by the Rhone gorge and valley through the Lake of Geneva to the great Rhone and the Mediterranean. Three incomparably fine days—September 17th, 18th, and 19th—atoned for three weeks of sunless cloud. One of them we spent in the high valley of Rosenlaui, where are hairy-lipped gentians and the blue-iced glacier, but of these I have not space to tell. Then the clouds and the rain resumed their odious domination, and we left Lucerne and its lakes invisible, overwhelmed in grey fog, and made for Paris.
October, 1910

THE PROBLEM OF THE GALLOPING HORSE

Until instantaneous photography was introduced, a little more than twenty-five years ago (by the discovery of the means of increasing the sensitiveness of a photographic plate), and gradually became familiar to everyone in the exhibitions known as the "biograph" or "cinematograph," the actual position of the legs in a galloping horse at any given fraction of a second was unknown. Anyone who has tried to "see" their position will agree that it cannot be done. Attempts had been made to make out what the movements and positions of the legs "must" be, by studying the hoof-marks in a soft track laid for the purpose. But the result was not satisfactory.

As everyone knows, the so-called "biograph" pictures are produced by an enormous series of consecutive instantaneous photographs taken on a continuous transparent flexible film or ribbon. The camera has a mechanism attached to it by which the sensitive film is jerked along so as to expose a length of two inches (the size of the picture given by the camera) for, say, one-thirtieth of a second without movement. The film is then jerked on and a second bit of two inches is brought into place for a thirtieth of a second and so on until a ribbon of some thousand pictures is obtained. The interval between each picture is usually also about one-thirtieth of a second, so that at least fifteen pictures are taken in every second of time, and according to the requirements of illumination and the rapidity of the movements of the men or animals photographed this number may be greatly increased. The film is developed, printed and fixed on a similar rolling mechanism and the pictures are thrown one by one by a powerful lantern on to a screen, and are jerked along at the same rate as that at which they were taken, and are magnified enormously. Animals and men in rapid movement, railway trains, the waves of the sea are thus photographed, and

27

when the serial pictures are thrown successively on the screen the result is that the eye detects no interval between the successive pictures—the figures appear as continuous moving objects. This is due to the fact that whilst the impression produced on the retina of the eye by each picture lasts for a tenth of a second (less with brighter light), the interval between the successive pictures is only one-thirtieth of a second, and accordingly the retinal impression has not gone or ceased before the next is there; hence there is no break in the series of retinal impressions, but continuity.[1]

It is this duration of the impression on the retina which prevents us from separating or "seeing distinctly" the successive phases of a horse's legs as he gallops by, and has led to the remarkable result that no artist has ever until twenty-five years ago represented correctly any one phase of the movement of the legs in a galloping horse, and it is doubtful whether that correctness is what the painter of a picture really ought to put on his canvas. If we examine the separate pictures of a galloping horse as taken on a cinematograph film, we have before us the actual record of the positions assumed by the legs at intervals of the thirtieth of a second (or whatever less interval and length of exposure may have been chosen), and it is simply astonishing to find how utterly different they are from what had been supposed. Twenty years ago Mr. Muybridge produced a number of these instantaneous photographs of moving animals—such as the horse in gallop, trot, canter, amble, walk, and jumping and bucking—also the dog running, birds of several kinds flying, camel, elephant, deer, and other animals in rapid movement. The animals were photographed on a track in front of a wall, marked out to show measured yards; the time was accurately recorded to show rate of movement and length of exposure, and of interval between successive pictures. By means of three cameras worked by electric shutter-openers, a side, a back, and a front view of the animal were taken simultaneously. Repeated photographs were obtained at intervals of a fraction of a second, giving a series of fifteen or twenty pictures of the moving animal. The length of exposure for each picture was one-fortieth of a second or less, and the interval between successive pictures was about the same. Muybridge's great difficulty had been to invent a shutter which would act rapidly enough. I have some of these pictures before me now (see Pl. I). They show that what has been drawn by artists and called the "flying gallop," in which the legs are fully extended and all the feet are off the ground, with the hind hoofs turned upwards, never occurs at all in the galloping horse, nor anything in the least like it. There is a fraction of a second when all four legs of the galloping horse are off the ground, but they are not then extended, but, on the contrary, are drawn, the hind ones forward and the front ones backward, under the horses' belly (see Pl. I, figs. 2 and 3). A model showing this actualinstantaneous attitude of the galloping horse has recently been placed in the Natural History Museum. When the hoofs touch the ground again after this instantaneous lifting and bending of the legs under the horse, the first to touch it is that of one of the hind legs (Pl. I, fig. 4), which is pushed very far forward, forming an acute angle with the body. The shock of the horse's impact on the ground is thus received by the hind leg, which reaches obliquely forward beneath the body like an elastic -spring. Since the instantaneous photographs have become generally known artists have ceased to represent the galloping horse in the curious stretched pose which used to be familiar to everyone in Herring's racing plates, with both fore and hind legs nearly horizontal, and the flat surface of the hind hoofs actually turned upwards! Indeed, as early as 1886 a French painter, M. Aimé Morot, availed himself of the information afforded by the then quite novel instantaneous photographs of the galloping horse, and exhibited a picture of the cavalry fight at Rezonville between the French and Germans, in which the old flying gallop does not appear, but the attitudes of the horses are those revealed by the new photographs. The picture is an epoch-making one, whether justifiable or not, and is now in the gallery of the Luxembourg. It must be noted that though Meissonier and others had succeeded in representing more truthfully than had been customary, other movements of the horse, such as "pacing," ambling, cantering, and trotting, yet in regard to them, also, more easily observed because less rapid, the instantaneous photograph served to correct erroneous conclusions.

Two very interesting questions arise in connection with the discovery by instantaneous photography of the actual positions successively taken up by the legs of a galloping horse. The first is one of historical and psychological importance, viz. why and when did artists adopt the false but generally accepted attitude of the "flying gallop"? The second is psychological and also physiological, viz. if we admit that the true instantaneous phases of the horse's gallop (or of any other very rapid movement of anything) cannot be seen separately by the human eye, but can only be separated by instantaneous photography, ought an artist to introduce into a picture, which is not intended to serve merely as a scientific diagram, an appearance which has no actual existence so far as his or other human eyes are concerned, viz. that of the actual pose assumed instantaneously and simultaneously by the four legs of the galloping horse? And further, if he ought not to do this, what ought he to do, on the supposition that his purpose is to convey to others the same impression of rapid movement which exists—not, be it observed, in his eye, or on the retina of that eye—but in his mind, as the result of attention and judgment?

The first of these questions has been answered by the great French authority on archæology and the history of art, M. Salomon Reinach,[2] whose writings are as lucid and terse as they are accurate, and solidly based on research. M. Reinach shows (and produces drawings to support his statement) that in Assyrian, Egyptian, Greek, Roman, mediæval, and modern art up to the end of the eighteenth century "the flying gallop" does not appear at all! The first example (so far as those schools are concerned) is an engraving by G. T. Stubbs in 1794 of a horse called "Baronet." The essential points about "the flying gallop" are that the fore-limbs are fully stretched forward, the hind limbs fully stretched backward, and that the flat surfaces of the hinder hoofs are facing upwards. After this engraving of 1794 the attitude introduced by Stubbs became generally adopted in English art to represent a galloping horse, and the French painter, Géricault, introduced it into France in 1821 in his celebrated picture, the "Derby d'Epsom," (see Pl. II, fig. 1) which is now in the Louvre.

Previously to this there had been three other conventional poses for the running horse in art, of which only the third (to be mentioned below) has any resemblance to a real pose, and that not one of rapid movement. We find: (1) The elongated or stretched-leg "prance" (French, "cabré allongé"), in which, whilst the front legs are off the ground, and all four legs are stretched nearly as much as in the flying gallop, there is this essential difference, viz. that the hoofs of the hind legs are firmly planted on the ground (see Pl. II, fig. 7). This pose is seen in a picture by the same artist (Stubbs) of two years' earlier date than that in which he introduced "the flying gallop." The "stretched-leg prance" is found in Egyptian works (Pl. II, fig. 8) of 580 b.c., and is a favourite pose to indicate the gallop, in ancient Assyrian as well as mediæval art, for instance, in the Bayeux tapestry (Pl. II, fig. 6). We find, further, (2) that the second pose made use of for this purpose is the "flexed-leg prance," in which all the four legs are flexed, so that the hind legs rest on the ground beneath the horse's body, whilst the forelegs "paw" the air. This is seen both in Egyptian, Greek, and Renaissance art (Leonardo, Raphael, and Velasquez). It is by no means so graceful or true to Nature as the next pose, but gives an impression of greater energy and rapidity. The third pose represents a kind of "prancing," and is seen on the frieze of the Parthenon (Pl. III, fig. 4), and in many subsequent Greek, Roman, and other works copied from or inspired by, this Greek original. One only of the hind legs is on the ground, and the animal's body is thrown up as though its advance were checked by the rein. It is called "the canter" by M. Reinach, but that term can only be applied to it when the axis of the body is horizontal and parallel to the surface of the ground.

The reader will perhaps now suppose that we must attribute the "flying gallop" to the original, if inaccurate genius of an eighteenth century English horse-painter. That, however, is not the case. M. Reinach has shown that it has a much more extraordinary history. It is neither more nor less than the fact that in the pre-Homeric art of Greece—that which is called "Mycenæan" (of which so much was made known by the discoveries of that wonderful man Schliemann when he dug up the citadel of Agamemnon)—the figures of animals, horses, deer, bulls (see the beautiful gold cups of Vaphio), dogs, lions, and griffins, in the exact conventional pose of "the flying gallop," are quite abundant! (See Pl. II, figs. 2, 3 and 4.) There was an absolute break in the tradition of art between the early gold-workers of Mykené (1800 to 1000 b.c.) and the Greeks of Homer's time (800 b.c.). Europe never received it, nor did the Assyrians nor the Egyptians. Thirty centuries and more separate the reappearance in Europe of the flying gallop—through Stubbs—from the only other European examples of it—the Mycenæan. What, then, had become of it, and how did it come to England? M. Reinach shows, by actual specimens of art-work, that the Mycenæan art tradition, and with it the "flying gallop," passed slowly through Asia Minor north eastwards to the Trans-caucasus (Koban, 500 b.c.), to Northern Persia, and thence by Southern Siberia to the Chinese Empire (Pl. III, fig. 2) as early as 150 b.c., and that the "flying gallop," so to speak, "flourished" there for centuries, and was transmitted by the Chinese artists to the Japanese, in whose drawings it is frequent (Pl. III, fig. 3). It was at last finally brought back to Europe, and to the extreme west of it, namely, England, by the importation in the eighteenth century into England of large numbers of Japanese works of art. It was a Japanese drawing (M. Reinach infers) which suggested to Stubbs the upturned hinder hoofs and the detachment from the ground of "the flying gallop" which he gave in his portrait of "Baronet," and so established that pose for a century in modern European art. This is a delightful tracing out of the wanderings of an artistic "convention," and the curious thing is that its chief importance is not that it has to do with the movements of the horse, but that it tends (as do other discoveries) to establish the gradual passage of pre-classical Mycenæan art across Central Asia to China and Japan by trade routes and human migrations which had no touch with later Greece nor with Assyria nor India.

How did the Mycenæans come to invent, or at any rate adopt, the convention of "the flying gallop," seeing that it does not truly represent either the fact or the appearance of a galloping horse? Though 20,000 years ago the earliest of all known artists, the wonderful cave-men of the Reindeer period, drew bison, boars, and deer in rapid running movement with consummate skill, they were (be it said to their credit!) innocent of the conventional pose of the "flying gallop." I base this statement on my own knowledge of their work. M. Reinach thinks that the "flying gallop" was devised as an intentional expression of energy in movement. I venture to hold the opinion that it was observed by the Mycenæans in the dog, in which Muybridge's photographs (now before me) demonstrate that it occurs regularly as an attitude of that animal's quickest pace or gallop (see fig. 5, Pl. II). It is easy to see the "flying gallop" in the case of the dog, since the dog does not travel so fast as the galloping horse, and can be more readily brought under accurate vision on account of its smaller size. The late Professor Marey (a great investigator of animal movement) appears to have denied that the dog exhibits the full stretch of both limbs with the pads of the hind-feet upturned, and all the feet free from the ground. He was mistaken, as Muybridge's photograph giving side and back view of a galloping fox-terrier amply demonstrates. It is quite in accordance with probability that the early Mycenæan artists, having seen how the dog gallops, erroneously proceeded to put the galloping horse, and all other animals which they wished "to make gallop," into the same position.

It appears, then, that the poses used by artists at different times and in different parts of the world to represent the "galloping" of the horse have no correspondence to any of the poses actually assumed by a galloping horse as now demonstrated by instantaneous photography. The "prancing" attitude of the horses of the frieze of the Parthenon was probably not intended to represent rapid movement at all. The "stretched-leg" pose and the "flex-leg" pose are, as a matter of fact, phases of "the jump," and are definitely recorded in Muybridge's instantaneous photographs of the jumping horse, but have no existence in "galloping" nor in any rapid running of the horse. They were probably adopted by the artists of Egypt, Assyria, Greece, and their successors in Europe as an expedient without conviction, to represent rapid movement, the true poses of which defied satisfactory reproduction. And it is also the fact that the "flying gallop," which appeared in Mycenæan art thirty-seven centuries ago, and then travelled by a "Scythian" route through Tartary to China, and came back to Europe at the end of the eighteenth century, is also—so far as it has any real representative in the action of the horse—only approached by a brief phase of the "jump." The poses of the horse in jumping are shown in the small figures taken from instantaneous photographs and reproduced in Fig. 6 of Pl. III. The "flying gallop" ("ventre a terre"), with all four legs stretched, and the under surface of the hind feet upturned, is really seen by us all every day in the dog, and is recorded in instantaneous photographs of that animal going at full speed. In fact, the gallop of the dog (and of some other small animals) is a series of jumps; the animal "bounds along." But this is a totally different thing from the gallop of the horse. It is probable that the dog's gallop was transferred, so to speak, to the horse by artists, and a certain justification for it was found in one of the attitudes of a jumping horse, which, however, never exhibits both the front and the hind legs simultaneously in so completely horizontal a position as they are made to take in the Mycenæan gold-work and the modern "racing plates."

How, then, we may now ask, ought an artist to represent a galloping horse?

Some critics say that he ought not to represent anything in such rapid action at all. But, putting that opinion aside, it is an interesting question as to what a painter should depict on his canvas in order to convey to others who look at it the state of mind, of impression, feeling, emotion, judgment, which a live, galloping horse produces in him. The scientific draughtsman would, of course, present to us a series of drawings exactly like the instantaneous photographs, his object being to show what "is," and not what the artist aims at, namely, what "appears," "seems," or (without pondering and analysis) "is thought to be." The painter, in his quality of artist, would be wrong to select any one of the dozen or more poses of the galloping horse published by Muybridge, each limited to the fortieth of a second, since no human eye can fix (as the photographic camera can) separate pictures following one another at the rate of twenty a second, each enduring one fortieth of a second, and each separated by an interval of a fortieth of a second from the next. All the phases which occur in any one-tenth of a second (only two, or possibly three of the Muybridge series shown in Pl. I) are, as it were, fused in our visual impression, because each picture lasts on the retina of the eye for one-tenth of a second, or (to put it more accurately) because the "impression" or condition of the retina produced by each picture persists or endures for the tenth of a second.

It may, perhaps, be suggested (and, indeed, has been), that it is the "blurred" or "fused" picture produced by the successive poses of the galloping horse's legs in one-tenth of a second that the painter ought to imitate on his canvas. In support of this notion we have the fact that the rapidly running wheels of a coach or of a gun-carriage (as in the pictures by Wouwerman) are represented by artists, not with the twelve or fourteen spokes which we know to be there—and would be photographed as separate things in an exposure of the fortieth of a second—but as a blurred haze of some fifty or more indistinct "spokes." In this case it undoubtedly results that the observer of the picture is satisfied and receives the mental impression or illusion of a rapid rotation of the wheel. I have tried the experiment with instantaneous photographs of the galloping horse, and I get three results: first, no combination of successive phases occupying one-tenth of a second gives anything resembling the "flying gallop" of the racing plates (the Mycenæan and Stubbsian pose), or any other conventional pose; second, no combination of successive instantaneous photographs limited to ten second gives any pose which satisfies the judgment and suggests a movement like the gallop; third, the combination which comes nearest to satisfying the judgment as being a natural appearance, but does not quite succeed in doing so, is one formed by the fusion of figs. 2 and 3 of Pl. I. This gives all four legs off the ground, drawn up or flexed beneath the horse's body, as in Morot's picture of the sabre-charge at Resonville.

The fact is that we have to take into consideration two other factors in the process, which we call "seeing," besides the duration of the retinal impression or excitation. These are, first, attention, and second, judgment. We are apt to think that "seeing" is a simple, straightforward sort of thing, whereas it is really a strangely complex and delusive process. "I did not see it, therefore it was not there," or "You must have seen it; it was right in front of you," are common assertions, and the belief that such assertions are justified leads to miscarriage of justice in courts of law. Yet everyone knows that he may stare out of the window of a railway carriage and have a long panorama pass before his eyes, or may walk along a crowded street and look his acquaintances in the face, and in neither case will he have "seen" or recognized anything, or be able to give an account of the scene that was pictured on the back of his eye. Attention, the direction of the mind to the sensation, is necessary; and it appears that it is very difficult (to some more than to others) to hold the attention alert, and to give it to the unexpected. In fact, to a very large extent we can only "see" (using the word to signify the ultimate mental condition) that which we are prepared to see or that which we expect to see. In the absence of such expectation, a very strongly illuminated or well-marked, outstanding object is far more readily "seen" than less marked objects. Accordingly, the outstretched legs of the galloping horse, now in front and now behind, are "seen," whilst the rest of the phases are not observed. Moreover, it is a fact that the swinging pendulum of a clock is "seen" at the extreme position of the swing on each side, and not in the intermediate space. This is because the image is formed very quickly, twice in the space where the bob of the pendulum is coming to the limit of its swing and is again returning on its course. For the same reason, the outstretched legs of the horse going up to their limit and at once returning give in very quick succession, near their extreme limit, an ascending and a descending phase which are not strictly but sensibly alike, and so doubly impress the retina, and obtain for the legs "attention" when in that extreme position. The choice of the attitude depicted by Morot is explained by the fact that, as is shown by its persistence through two successive pictures (figs. 2 and 3 of Pl. I), this pose must produce a more continuous impression on the retina than any other of the attitudes shown, since none of them endure through two successive pictures.

The mental process of attention results in a certain duration or memory of the mental condition which is a distinct thing from the primary retinal impression, and leads to the ignoring or mental obliteration of an instantaneous interval separating two phases of the position of moving legs which have strongly "arrested the attention." Hence, it seems that the most forward pose of the galloping horse's front legs and the most backward pose of its hind legs—though far from simultaneous, even in the slow changing retinal impressions—may be mentally combined by "the arrest of attention," and that the artist really ought to present his picture of the galloping horse with those two poses combined (although as a matter of scientific truth they do not occur simultaneously) in order that he may produce by his painted piece of canvas, as nearly as he can, the mental result which we call "seeing" a horse gallop. This combination of the front half of one figure with the hinder half of another so as to give in each case the extreme phase of extension of the legs I have made in Pl. I, fig. 12.

But there is, further, in all "seeing" before even a mental result of attention to the retinal picture is, as it were, "passed," admitted and registered as "a thing seen," the further operation of rapid criticism or judgment, brief though it be. We are always unconsciously forming lightning-like judgments by the use of our eyes, rejecting the improbable, and (as we consider) preposterous, and accepting and therefore "seeing" what our judgment approves even when it is not there! We accept as "a thing seen" a wheel buzzing round with something like fifty spokes—but we cannot accept a horse with eight or sixteen legs! The four-leggedness of a horse is too dominant a prejudice for us to accept a horse with several indistinct blurred legs as representing what we see when the horse gallops. The mind revolts at such a presentation, though it is true, and the whole scheme and composition of the artist is perverted or fails to gain attention and to exercise its charm—by the unwelcome presence in his picture of the revolting truth. It is the consideration of facts of this kind which enables us to understand the origin and importance of what are called "conventions" in pictorial or glyptic art. The artist is, in fact, operating by means of his painted canvas or moulded clay upon a queer, prejudiced, ill-seeing, dull, living creature—his brother-man. In order to give if possible to that brother, by means of a painted sheet, some or all of the delights, emotions, suggestions, perceptions of beauty, and so on, which he himself has experienced in contemplating a real scene, the artist has to present that scene, not as it really is, nor even as he thinks it really is, but in such a way that his canvas shall appeal to his brother's attention and judgment with the same emotional and intellectual result as the scene itself produced in him. Therefore he must not aim at accuracy of reproduction of natural fact nor even of visual fact, but at the transference to another mind of his own mental condition—his inner judgment as to "things seen"—by means of necessarily imperfect pictorial mimicry. He must therefore avoid startling or abnormal truthfulness of observation of the unessential and even more strictly must he refuse to make his picture a scientific diagram demonstrating what "is" rather than what is "seen" or is "thought to have been seen."

On these grounds I find that the most satisfactory pictures of the galloping horse are those which combine a phase of the movement of the front legs with a phase of the movement of the hind legs, not simultaneous in actual occurrence, but following one another. It is for the artist to select the combination best suited to producing the mental result aimed at. Some of the Chinese and Japanese representations of the galloping horse and some of their European imitations (but not all—certainly not that of Stubbs, of the Epsom Derby of Géricault, and the racing plates) seem to me to be eminently satisfactory and successful in this respect. In the pictures to which I allude (Pl. III, figs. 3 and 5) all the legs are off the ground; the front legs are advanced, but one or both may be more or less flexed, whilst the hind legs, though directed backwards with upturned hoofs, are not nearly horizontal (as they actually are in the galloping dog), but show the moderate extension which really occurs in the horse, and is recorded by instantaneous photography. This pose, favoured by many European and Japanese artists, can be obtained by uniting the outstretched hind legs of fig. 9 of the Muybridge series (Pl. I), with the outstretched forelegs of fig. 6, as shown in Pl. I, fig. 12, or by uniting the hind legs of fig. 10 with the forelegs of fig. 4 as shown in Pl. III, fig. 1.

With regard to the representation of other "gaits" of the horse than that of the rapid gallop—such as canter, trot, amble, rack, and walk—I have no doubt that instantaneous photography can (and in practice does) furnish the painter with perfectly correct and at the same time useful and satisfactory poses of the horse's limbs. These, though of longer duration than the poses of the gallop, can only be correctly estimated by the eye with great difficulty, and only sketched by artists of exceptional skill and patience. The movement of the wings of birds in flight has been very successfully analysed by instantaneous photography. Some of the poses revealed must familiarise the public with what can be, and, in fact, has been, observed in the case of large sea-birds, by the unassisted eye, and has been represented in pictures by the more careful observers of nature among modern painters. A large sea-bird sailing along with apparently motionless wings has been photographed in the act of giving a single stroke so rapid as to escape observation by the eye.

An interesting question in regard to the movements of the horse is that as to how far any known "pace" is natural to that animal, and how far it has been acquired by training and is, in a sense, artificial. We know so little of the wild horse, and of the more abundant wild asses and zebras, that it is difficult to say anything precise on this question. There is only one region in which the true original wild horse of the northern part of Asia and Europe still exists. That is the Gobi Desert, in Central Asia. This horse is known as Prevalsky's wild horse, in honour of the Russian traveller who discovered it. Live specimens are now to be seen in the Zoological Gardens and elsewhere. It closely resembles the drawings of horses made by the palæolithic Cromagnard cave-men. A century ago a wild horse, probably of the same race as this, inhabited the Kirghiz Steppes, and was known as the Tarpan: it is now extinct. The more southern Arabian horse is not known in the wild state, whilst the wild horses of America are descendants of domesticated European horses which have "run wild." I do not know of any studies of the movements of the true wild horse, nor of those of wild asses and zebras, carried out by the aid of instantaneous photography. It would be interesting to know whether untaught wild "equines" would fall naturally into the gaits known as "the amble" and "the rack," or whether the walk, the trot, and the gallop are their only natural gaits.

The amble, in which the fore and hind leg on the same side are advanced simultaneously, is a natural gait of the elephant, the fastest Muybridge could get from that great beast. He made a menagerie elephant amble at the rate of a mile in seven minutes. The only other animal known to habitually exhibit "the amble" is the giraffe. It is often exhibited by the giraffes in the Zoological Gardens in London, but has not, I believe, been recorded by a series of instantaneous photographs. When going at full speed over the grass wilds of Central Africa the giraffe exhibits a gait more like the galloping of deer and antelopes, and carries the long neck horizontally. No complete study of the "gaits" of large animals other than the horse has been made, since menagerie specimens and menagerie conditions are not satisfactory for the purpose, and, unfortunately, it has not been possible as yet to take series of photographs of them in their wild conditions.

The electric spark furnishes a most important means of taking instantaneous photographs, but the operator must perform in the dark. An electric spark can be obtained which lasts only the one two-thousandth of a second, and by its use as the sole illuminating agent we can get a photograph of a phase of movement lasting only that excessively short space of time, or, if we please, a succession of such phases by using a succession of sparks. Thus, a rifle bullet is readily photographed while in flight with scarcely perceptible distortion. A wheel revolving many hundred times a second can thus be photographed, and appears to be stationary. Dr. Schillings has applied this method to the photography of wild animals by night in the forests of tropical Africa, and has published an interesting book giving his photographic results. In order to take these pictures the track followed by certain animals has to be detected, and then a thread is stretched "breast-high" across the track, so that the animal coming along it by night shall pull the thread. Immediately the thread is pulled it sets an electric contact in action. There is a brief flash of one two-thousandth of a second, and a picture is taken by a camera previously fixed, out of harm's way, so as to focus the area where the thread was stretched.

Dr. Schillings obtained some very remarkable photographs of "the night life of the forest" in this way—lions and leopards advancing on their prey were suddenly revealed, and the helpless antelope or other victim was shown crouching in the dark, or making a desperate effort to escape.

The electric-spark method was applied by a friend of mine to demonstrate the movements by which a kitten falling backwards from a table succeeds in turning itself so as to alight on its feet. During a fall of less than 3 feet he obtained five successive spark-pictures of the kitten, which, I beg it may be clearly understood, was a pet kitten, and was neither frightened nor hurt by the proceedings.

Instantaneous photographs, whether obtained by the use of an electric spark as a means of illumination, or by the less rapid method of a spring shutter working in combination with a sensitive film, which is jerked along so as to be exposed when the shutter is open and travel when it is shut, has been applied to the analysis of other movements than those I have mentioned, and has yet to be applied to many more, such as the crawling of insects and millipedes, and the beautiful rippling movement of the legs and body by which many marine worms swim. It has been extensively used in the study of human locomotion, and of the successive poses of the arms and legs in various athletic exercises, and in such games as baseball and golf. A first-rate fencer of my acquaintance had a five-minutes' film of himself taken when fencing, giving 10,000 consecutive poses. He wished to see exactly what movements he made, and to ascertain by this minute examination any error or want of grace in his action, in order to avoid it. An unexpected picture is obtained when a man or woman is thus "biographed" whilst walking rapidly, and suddenly turns to the right or left. A fraction of a second occurs when the toes of the two feet are directed towards one another (that is to say, are "turned in"), as one of the legs swings round in the break-off to right or left. This instantaneous phase is very awkward and ugly in appearance. It is never pictured by artists, although regularly occurring, and seems to have been as little known before instantaneous photography was introduced as were most of the phases of the horse's gallop. The positions assumed when in the air by a high-jump athlete are almost incredible as revealed by the camera. He appears to be sitting in a most uncomfortable way on the rope over which he is projecting himself.

A very fine attitude is fixed for the artist in one of Muybridge's instantaneous series of the "bowler"—the cricket "bowler." The up-lifted right arm, the curve outwards of the whole figure on the right side, and the free hang of the right leg make a most effective pose for a sculptor to reproduce. Among the most remarkable results obtained in Muybridge's series are the stages of the growth or development of strong "expression" in the face. The anxiety in the face of the baseball batsman as he awaits the ball is painful; as he hits at the ball his expression is one of savage ferocity, and in a fraction of a second this gives place to a dawning smile, which as we pass along two or three later "instantanèes" develops into a broad grin of satisfaction. Another genuine study of expression both of face and gesture and movement is given in the series where a pailful of cold water is unexpectedly poured over the back of a bather seated in a sitz bath—

astonishment, dismay, anger, eagerness to escape, and the reaction to shock are all clearly shown. Darwin's studies on "the expression of the emotions" would have been greatly assisted by such analysis, and the subject might even now be developed by the use of serial instantaneous records obtained by photography. It may be useful to those interested in this subject to know that copies of Muybridge's large series of instantaneous photographs[3] of animal and human subjects in movement are preserved both in the library of the Royal Academy of Arts in London and in the Radcliffe Library at Oxford. I may also mention the extremely valuable series of instantaneous photographs of living bacteria, blood-parasites and infusoria produced by MM. Pathé, and the series of fishes and various invertebrates (including the curious caterpillar-like Peripatus) taken by Mr. Martin Duncan.

The representation of the moon in pictures of the ordinary size (some three feet long by two in height) is a case in which the artist habitually—one may almost say invariably—departs greatly from scientific truth, and it is a question as to whether he is justified in what he does. Take first the case of the low-lying moon near the horizon as contrasted with the high moon. Everyone knows that the moon (and the sun[4] also) appears to be much bigger when it is low than when it is high. Everyone who has not looked into the matter closely is prepared to maintain that the luminous disc in the sky—whether of moon or of sun—not merely seems to, but actually does, occupy a bigger space when it is low down near the horizon than when it is high up, more nearly overhead. Of course, no one nowadays imagines that the moon or the sun swells as it sinks or diminishes in volume as it rises. Those who think about it at all, say that the greater length of atmosphere through which one sees the low sun or moon, as compared with the high, magnifies the disc as a lens might do. This, however, is not the case. If we take a photograph of the moon when low and another with the same instrument and the same focus when it is high, we find that the celestial disc produces on the plate (as it does on our eyes) a picture-disc of practically the same size in both positions. In fact, the high moon or sun produces a picture-disc of a little larger size than the low moon or sun. I have here reproduced (Pl. IV) a photograph, published by M. Flammarion, in which the moon has been allowed to print itself on a photographic plate exposed during the time the moon was rising, and it is seen that the track of the moon has not diminished in width as it rose higher and higher. No one will readily believe this, yet it is a demonstrable fact. Astronomers have made accurate measurements which show that there is no diminution of the disc under these circumstances, but a slight increase—since the moon is a very little nearer to us when overhead than when we see it across the horizon.

If we put a piece of glass coated with a thin layer of water-colour paint into a frame, and then make a peep-hole in a board which we fix upright between us and the upright piece of framed glass, we can keep the framed glass steady (let us suppose it to be part of the window of a room), and then we can move the peep-hole board back from it into the room to measured distances. At a distance of one and a half feet from the framed glass, which is that at which an artist usually has his eye from his canvas or paper, we can trace on the smeared or tinted piece of glass the outlines of things seen through it exactly as they fill up the area of the glass—men, houses, trees, the moon. The moon's disc (and the same is true of the sun) is found always to occupy a space on the glass which is 1/115th of the distance of the eye from the framed glass plate. When the eye-to-frame distance is eighteen inches, the diameter of the disc of the moon on the smeared glass will occupy exactly 1/115th of eighteen inches, which is between one-sixth and one-seventh of an inch. Similarly if the peep-hole is at nine and a half feet or 114 inches from the framed glass (which stands for us as the equivalent of an artist's picture) the moon will occupy almost exactly one inch in diameter—the size of a halfpenny. With such a simple apparatus of peep-hole and smeared glass in an upright frame, it is easy to mark off the size covered by the moon (or sun), whether low or high, on the smeared glass, and it is found never to vary whether high or low—so long as the same "eye-to-frame" or "peep-hole" distance is preserved. That seems to be an important fact for painters of sun-sets and moon-rises. But what do they do? They never give the right size (namely one-sixth of an inch) which corresponds to an eye-to-frame distance of eighteen inches. They give to a high moon, if they are very careful, a quarter of an inch for diameter. This means that the observer is about two and a half feet, or thirty inches from the picture—nearly twice what the artist's eye really is as he paints. And then—if painting a moon-rise or sunset—they suddenly pretend to go to a distance of nine and a half feet from the picture and make the moon an inch across because it is low down, or even give the moon two inches in diameter, which would mean that they (and those who look at the picture when hung up for view) are observing at nineteen feet distance from the front plane or frame of the picture. They do not alter the other features in the picture to suit this change of distance of the eye from the frame and there is no warning given. Certainly there is no obvious and necessary reason for treating a picture containing a high moon as though you were three feet from the front plane of the scene presented, and a low moon as though you were twenty feet from that plane! The confusion which may result in the representation of other objects when these changes of eye-to-frame distance are made is shown by the following simple facts. According to the simple laws of perspective, if the eye is at thirty inches from the picture-plane or frame (as declared by a moon drawn of a little more than a quarter of an inch broad), a post or a man six feet high drawn on the canvas as three inches high absolutely and definitely means that that man or post is sixty feet away from the observer inside the picture. The height of the represented object is the same fraction of the real object as the eye-to-frame distance is of the distance of the observer to the real object. If by a two-inch moon the artist has thrown you back from the front plane of the scene to a distance of nineteen feet, then the six-foot post or man drawn as three inches high definitely asserts that it or he is 456 feet distant within the picture. So, too, if the church tower which cuts the moon is really sixty feet high and is drawn of two inches vertical measure in the picture, it is an assertion—when the moon is represented one quarter of an inch broad—that the church tower is 290 yards, or a sixth of a mile distant. If, on the other hand, other things remaining the same, the moon is drawn two inches in diameter, the church tower is now asserted to be eight times as far off, or about a mile and a third. Very generally these facts are not considered by painters. They represent the low moon (or sun) big because the erroneous mental impression is common to all of us that it is big—that is, bigger, much bigger, than the high moon or sun, and they do not follow out the consequences in perspective of the pictorial increase of the moon's apparent diameter.

If we could ascertain why it is that the low moon produces a false impression of being bigger—as a mere disc in the scene—than does the high moon, we might be able to discover how an artist could produce, as Nature does, an impression or belief in its greater size whilst keeping it all the time to its proper size. The explanation of the illusion as to the increased size of the sun's or moon's disc when low, given by M. Flammarion and other astronomers, is that the low sun or moon is unconsciously judged by us as an object at a greater distance than the high moon or sun. This is due to the long vista of arching clouds above and of stretching landscape or sea below when the sun or moon is looked at as it appears on or near the horizon. The illusion is aided by the dulness of the low moon and the brightness (supposed nearness) of the high moon. Being judged of (unconsciously) as further off than the high moon, the low moon is estimated as of larger size although of the same size. This is, I believe, the correct explanation of the illusion. When one gazes upwards to the sky, a small insect slowly flying across the line of sight sometimes is "judged of" as a huge bird—an eagle or a vulture—since we refer it to a distance at which birds fly and not to the shorter distance to which insects approach us. It seems that it would be possible for the painter, by carefully studying actual natural facts and introducing their presentation into his picture, to produce the impression of greater distance, and therefore of size, into a quarter-inch moon placed near the horizon. He is not compelled for want of other means to "cut the difficulty" and paint a falsely inflated moon which shall brutally and by measurement call up the illusion of increased size. I reproduce here (Pl. V) an interesting drawing which shows how such illusions of size can be produced. It is none the worse for my purpose because it is an advertisement by the well-known firm who have kindly lent it to me. The three figures represented in black are all of the same height, yet the furthest one appears to be much taller and bigger altogether than the middle one, and the middle one than the nearest. This result is obtained by suggesting distance as separating the right-hand figure from us, whilst giving it exactly the same height as the others. This seems to me to be a simple case of an illusion of increased size produced by a suggestion of increased distance when all the time there is equality in size—as in the case of the moon on the horizon compared with the moon overhead. It would be interesting to see an attempt on the part of a competent painter to produce in this way (which is, I believe, Nature's way) the illusion of increased size in a low-lying moon without really increasing the visual size of his painted moon as compared with one in another picture (to be painted by him) representing the moon bright, clear and small, overhead.

The theatrical scene-painter has another kind of difficulty with the low moon and the setting sun. He can never be right for more than one row of seats—one distance—in the theatre. Here there is no peep-hole, no frame or picture-plane. The observer is in the picture. If the moon is represented by an illuminated disc of one foot in diameter, it will, when looked at at a distance of 115 feet, have the same visual size as the moon itself, but if your seat is nearer the scene it will look too large, if further off it will look too small. There is no getting over this difficulty, as the standard of actual Nature is set up on the stage by the men and women appearing on it at a known distance. It used to be asked in classical times by ingenious puzzle-makers—"What is the size of the moon?" A true answer to that question would be "that of a plate a foot in diameter seen at a distance of a hundred and fifteen feet."

To a large extent the painter, like other artists, has to produce things which do not shock common opinion and experience, and must even consciously concede to that necessity, and make the sacrifice of objective truth, in order to secure attention for his higher appeal to the sense of beauty, to emotion, and sentiment. Approved departures by the artist from scientific truth are those which are deliberately made in order to give emphasis—as, for instance, in the huge, but tender hand of the man in the emotional masterpiece, "Le Baiser," by the great sculptor Rodin. Another departure from objective truth which is justified, is seen in Troyon's picture in the Louvre, where the false drawing and exaggerated size of the leg of a calf advancing towards the observer suggest, and almost give the illusion of, movement.

But it can hardly be maintained that any and all the liberties which a painter or a whole school of painters choose to take with fact in their presentation of Nature—are beyond criticism. It is possible for a landscape painter to improve in his treatment of the moon by better observation and increased knowledge—just as other painters have learnt not to introduce into their pictures the sort of wooden rocking-horse to stand for a beautiful living animal, which satisfied Velasquez, Carl Vernet and the ancient Egyptians.

[1] See note on page 46.

[2] "La Representation du Galop dans l'art ancien et moderne," 'Revue Archeologique,' vol. xxxvi et seq., 1900.

[3] A word is needed in amplification of what was said on p. 26 as to the blending of successive images produced on the retina of the eye by the bioscope or cinematograph or by the old "wheel of life." The point which is of importance is not the length of time during which the stimulation of the retina caused by an image endures—becoming weaker and weaker as fractions of a second pass—but it is this: How long will a stimulus last in undiminished brightness? How soon must it be followed by another stimulus (another image) so that there may be fusion or continuity, the one succeeding the other before the earlier has had time, not to disappear, but to decline. If it has had time to decline in intensity, the appearance of flickering results. That is what the cinematographer has to avoid. It is found that a quicker succession—a shorter interval—is necessary with strong light than with weaker light in order to produce continuity. With a faint light the interval may be as great as one-tenth of a second; with a strong light it must not exceed one-thirtieth (or with still stronger light, one-sixtieth) of a second. With the stronger light there is a more rapid and a greater loss of the initial intensity of the impression or effect of stimulus, and though each successive effect remains as long, or longer, in dwindling intensity, you get want of continuity, or "flicker."

[4] What we may call the "visual size" of the sun happens to be owing to its far greater size and its far greater distance from us—very nearly the same as that of the moon—and is subject to the same numerical law of apparent diameter, viz. a disc of any given measurement in diameter will cover it exactly when held at a distance from the eye which is 115 times that measurement.

THE JEWEL IN THE TOAD'S HEAD

To what jewel or precious stone was Shakespeare alluding when he makes the exiled Duke in "As You Like It" (after praising his rough life in the forest of Arden, and declaring that adversity has its compensations), exclaim:

"The toad, ugly and venomous,
Wears yet a precious jewel in his head"?

No doubt the unprejudiced reader supposes when he reads this passage that there is some stone or stone-like body in the head of the toad which has a special beauty, or else was believed to possess magical or medicinal properties. And it is probable that Shakespeare himself did suppose that such a stone existed. As a matter of fact there is no stone or "jewel" of any kind in the head of the common toad nor of any species of toad—common or rare. This is a simple and certain result of the careful examination of the heads of innumerable toads, and is not merely "common knowledge," but actually the last word of the scientific expert. In these days of "nature study" writers familiar with toads and frogs and kindred beasts have puzzled over Shakespeare's words, and suggested that he was really referring to the beautiful eyes of the toad, which are like gems in colour and brilliance.

This, however, is not the case. Shakespeare himself was simply making use of what was considered to be "common knowledge" in his day when he made the Duke compare adversity to the toad with a magic jewel in its head commonly known as "a toad-stone," although that "common knowledge" was really not knowledge at all, but—like an enormous mass of the accepted current statements in those times, about animals, plants and stones—was an absolutely baseless invention. Such baseless beliefs were due to the perfectly innocent but reckless habit of mankind, throughout

45

long ages, of exaggerating and building up marvellous narrations on the one hand, and on the other hand of believing without any sufficient inquiry, and with delight and enthusiasm, such marvellous narrations set down by others. Each writer or "gossip" concerning the wonders of unexplored nature, consciously or unconsciously, added a little to the story as received by him, and so the authoritative statements as to marvels grew more and more astonishing and interesting.

It was not until the time of Shakespeare himself that another spirit began to assert itself—namely, that of asking whether a prevalent belief or tradition is actually a true statement of fact. Men proceeded to test the belief by an examination of the thing in question, and not by merely adducing the assertions of "the learned so-and-so," or of "the ingenious Mr. Dash." This spirit of inquiry actually existed in a fairly active state among the more cultivated of the ancient Greeks. Aristotle (who flourished about 350 b.c.), though he could not free himself altogether from the primitive tendency to accept the marvellous as true because it is marvellous and without regard to its probability—in fact because of its improbability—yet on the whole showed a determination to investigate, and to see things for himself, and left in his writings an immense series of first-rate original observations. He had far more of the modern scientific spirit than had the innumerable credulous writers of Western Europe who lived fifteen hundred to two thousand years after him. Even that delightful person Herodotus, who preceded Aristotle by a hundred years, occasionally took the trouble to inquire into some of the wonders he heard of on his travels, and is careful to say now and then that he does not believe what he heard. But the mediæval-makers of "bestiaries," herbals, and treatises on stones, which were collections of every possible fancy and "old-wife's tale," about animals, plants, and minerals, mixed up with Greek and Arabic legends and the mystical, medical lore of the "Physiologus"—that Byzantine cyclopædia of "wisdom while you wait"—deliberately discarded all attempt to set down the truth; they simply gave that up as a bad job, and recorded every strange story, property and "application" (as they termed it) of natural objects with solemn assurance, adding a bit of their own invention to the gathered and growing mass of preposterous misunderstanding and superstition.

In the seventeenth century the opposition to this method of omnivorous credulity (which even to-day, in spite of all our "progress," flourishes among both the rich and the poor) crystallised in the purpose of the Royal Society of London for the Improvement of Natural Knowledge—whose motto was, and is "Nullius in verba" (that is, "We swear by no man's words"), and whose original first rule, to be observed at its meetings, was that no one should discourse of his opinions or narrate a marvel, but that any member who wished to address the society should "bring in," that is to say, "exhibit" an experiment or an actual specimen. A new spirit, the "scientific" spirit, gave rise to and was nourished by this and similar societies of learned men. As a consequence the absurdities and the cruel and injurious beliefs in witchcraft, astrology, and baseless legend, melted away like clouds before the rising sun. In the place of the mad nightmare of fantastic ignorance, there grew up the solid body of unassailable knowledge of Nature and of man which we call "science"—a growth which made such prodigious strides in the last century that we now may truly be said to live in the presence of a new heaven and a new earth!

It was, then, a real "stone," called the toad-stone, to which Shakespeare alluded. It is mentioned in various old treatises concerning the magical and medicinal properties of gems and stones under its Latin name, "Bufonius lapis," and was also called Borax, Nosa, Crapondinus, Crapaudina, Chelonitis, and Batrachites. It was also called Grateriano and Garatronius, after a gentleman named Gratterus, who in 1473 found a very large one, reputed to have marvellous power. In 1657, in the "translation by a person of quality" of the "Thaumatographia" of a Polish physician named Jonstonus, we find written of it: "Toads produce a stone, with their own image sometimes. It hath very great force against malignant tumours that are venomous. They are used to heat it in a bag, and to lay it hot, without anything between, to the naked body, and to rub the affected place with it. They say it prevails against inchantments of witches, especially for women and children bewitched. So soon as you apply it to one bewitched it sweats many drops. In the plague it is laid to the heart to strengthen it." Another physician of the same period (see "Notes and Queries," fourth series, vol. vii, 1871, p. 540) appears to be affected by the new spirit of inquiry, for he relates the old traditions about the stone and how he tested them. He says it was reported that the stone could be cut out of the toad's head. (In the book called "Hortus Sanitatis," dated 1490, there is a picture, here reproduced [Fig. 4], of a gentleman performing this operation successfully on a gigantic toad.) Our sceptical physician, however, goes on to say that it was commonly believed that these stones are thrown out of the mouth by old toads (probably the tongue was mistaken for the stone), and that if toads are placed on a piece of red cloth they will eject their "toad-stones," but rapidly swallow them again before one can seize the precious gem! He says that when he was a boy he procured an aged toad and placed it on a red cloth in order to obtain possession of "the stone." He sat watching the toad all night, but the toad did not eject anything. "Since that time," he says, "I have always regarded as humbug ('badineries') all that they relate of the toad-stone and of its origin." He then describes the actual stone which passes as the toad-stone, or "Bufonius lapis," and says that it is also called batrachite, or brontia, or ombria. His description exactly corresponds with the "toad-stones" which are well known at the present day in collections of old rings.

I have examined twelve of these rings in the British Museum, through the kindness of Sir Charles Read, P.S.A., the Keeper of Mediæval Antiquities, and four in the Ashmolean Museum at Oxford. Two of these are of chalcedony, with a figure of a toad roughly carved on the stone, and are of a character and origin different from the others. The others, which are the true and recognised "toad-stones" or "Bufonius lapis," are circular, slightly convex "stones," of a drab colour, with a smooth enamel-like surface. They are plate-like discs, being of thin substance and concave on the lower surface, which has an upstanding rim. I recognised them at once as the palatal teeth of a fossil fish called "Lepidotus," common in our own oolitic and wealden strata, and in rocks of that age all over the world. I give in Fig. 5 a drawing of a complete set of these teeth and of a single one detached. They were white and colourless in life, but are stained of various colours according to the nature of the rock in which they were embedded. A drab colour like that of the skin of the common toad is given to them by the iron salts present in many oolitic rocks; those found in the wealden of the Isle of Wight are black. That the "toad-stones" mounted in ancient rings are really the teeth of a fish has been already recorded by the Rev. R. H. Newell ("The Zoology of the English Poets," 1845), but he seems to be mistaken in identifying them with those of the wolf-fish (Anarrhicas). They undoubtedly are the palatal teeth of the fossil extinct ganoid fish Lepidotus. Before leaving the queer inventions and assertions of the old writers about these fossil teeth, which they declared to be taken out of the toad's head, let me quote one delightful passage from a contemporary of Shakespeare (Lupton: "A thousand notable things of sundry sortes. Whereof some are wonderful, some strange, some pleasant, divers necessary, a great sort profitable, and many very precious," London, 1595). "You shall know," he says, "whether the Toadstone called 'crapaudina' be the right and perfect stone or not. Hold the stone before a toad, so that he may see it. And if it be a right and true stone, the toad will leap towards it and make as though he would snatch it from you; he envieth so much that a man should have that stone. This was credibly told Mizaldus for truth by one of the French King's physicians, which affirmed that he did see the trial thereof."
We have thus before us the actual things called toad-stones, and believed by Shakespeare and his contemporaries to be found in the head of the toad. How did it come about that these pretty little button-like, drab-coloured fossil teeth were given such an erroneous history? This question was answered by the late Rev. C. W. King, Fellow of Trinity College, Cambridge, in his book on "Antique Gems" (London, 1860). He says, "I am not aware if any substance of a stony nature is ever now discovered within the head or body of the toad. Probably the whole story originated in

the name Batrachites (frog-stone or toad-stone), given in Pliny to a gem brought from Coptos, and so called from its resemblance to that animal in colour." We have not, it must be noted, any specimens of the toad-stone at the present day actually known to have been brought from Coptos. It is quite possible that the fossil fish-tooth was substituted ages ago for Pliny's Batrachites, and was never found at Coptos at all! Whether that is so or not, the fact is that Pliny never said it came out of a toad, but merely that it was of the colour of a toad.

The Pliny referred to is Pliny the Elder, the celebrated Roman naturalist who wrote a great treatise on natural history, which we still possess, and died in a.d. 79 whilst visiting the eruption of Vesuvius. He says nothing of the Batrachites being found inside the toad, nor does he mention its medicinal virtues. The name alone—simply the name "Batrachites," the Greek for toad-stone—was sufficient to lead the fertile imagination of the mediæval doctors to invent all the other particulars! It is a case precisely similar to that of the old lady who was credited with having vomited "three black crows." When the report was traced step by step to its source it was found that her nurse had stated that she vomited something as black as a crow!

The belief in the existence of a stone of magical properties in the head of the toad is only one of the many instances of beliefs of a closely similar kind which were accepted by Pliny (although he records no such belief as to the toad-stone), and were passed on from his treatise on natural history in a more or less muddled form to the middle ages, and so to our own time by later writers. Thus Pliny cites, as stones possessing magical properties, the "Bronte" found in the head of the tortoise, the Cinædia in the head of a fish of that name, the Chelonites, a grass-green stone found in a swallow's belly, the Draconites, which must be cut out of the head of a live serpent, the Hyænia from the eye of the Hyæna, and the Saurites from the bowels of a green lizard. All these and the Echites, or viper-stone, were credited with extraordinary magical virtues, and many of the assertions of later writers about the toad-stone are clearly due to their having calmly transferred the marvellous stories about other imaginary stones to the imaginary toad-stone. The only stone in the above list which has a real existence is that in the fish's head. Fish have a pair of beautiful translucent stones in their heads—the ear-stones or otoliths—by the laminated structure of which we can now determine the age of a fish just as a tree's age is told by the annual rings of growth in the wood of its stem. The fresh-water crayfish has a very curious pair of opaque stones (concretions of carbonate and phosphate of lime) formed in its gizzard as a normal and regular thing. They are familiar to every student who dissects a crayfish, and I am told that in Germany to-day, as in old times also, the "krebstein" is regarded by the country-folk as possessed of medicinal and magical properties. I am not able, on the present occasion, to trace out the possible origin of all the stories and beliefs about stones occurring within animals. They are more numerous than those cited by Pliny; they exist in every race and every civilization and refer to a large variety of animals. Probably many of these beliefs date from prehistoric times. In the East the most celebrated of these stones, since the period of Arabic civilisation, is called a bezoar-stone, "Bezoar" is the Persian word for "antidote," and does not apply only to a stone. The true and original "bezoar-stone" of the East is a concretion found in the intestine of the Persian wild goat. Those which I have seen are usually of the size and shape of a pigeon's egg and of a fine mahogany colour, with a smooth, polished surface. The Persian goat's bezoar-stone is found, on chemical analysis, to consist of "ellagic acid," an acid allied to gallic acid, the vegetable astringent product which occurs in oak-galls used until lately in the manufacture of ink. The bezoar-stone is probably a concretion formed in the intestine from some of the undigested portions of the goat's food. Such concretions are not uncommon, and occur even in man. "Bezoar-stones" are obtained in the East from deer, antelopes, and even monkeys, as well as goats, and must have a different chemical nature in each case. Minute scrapings from these stones are used in the East as medicine, and their chemical qualities render their use not altogether absurd, though they probably have not any really valuable action. It is probable that their use had a later origin than that of the "stones" connected with magic and witchcraft. Sixteenth century writers, ever ready to invent a history when their knowledge was defective, declared the bezoar-stone to be formed by the inspissated tears of the deer or of the gazelle—the "gum" which Hamlet remarked in aged examples of the human species.

The substance called "ambergris" (grey amber), valued to-day as a perfume, is a fæcal concretion similar to a bezoar-stone. It is formed in the intestine of the sperm-whale, and contains fragments of the hard parts of cuttle-fishes, which are the food of these whales. "Hair-balls" are formed in the intestines of various large vegetarian animals—and occasionally stony concretions of various chemical composition are formed in the urinary bladder of various animals, as well as of man. The "eagle-stone" is also a concretion to which magical properties were ascribed. I have seen a specimen, but do not know its history and origin. Glass beads found in prehistoric burial-places are called by old writers "adders' eggs," and "adder-stones," and were said (it is improbable that one should say "believed") to hatch out young adders when incubated with sufficiently silly ceremonies and observances. A celebrated "stone" of medicinal reputation in the East is the "goa-stone." This is a purely artificial product—a mass of the size and shape of a large egg, consisting of some very fine and soft powder like fullers'-earth, sweetly scented, and overlaid with gold-leaf. A very little is rubbed off, mixed with water, and swallowed, as a remedy for many diseases. The deep connection of medicine with magic throwing light on the strange application of stones and hairs, bones and skins, by imaginative mankind, in all ages and places, is exhibited in the common practice of writing with ink a sentence of the Koran (or other sacred words) on a tablet, washing off the ink and making the patient swallow the water in which the sacred phrase has been thus dissolved! How convenient it would be were it possible thus to impart knowledge, virtue, and health to suffering humanity!

A good example of one of the ways in which magical properties become attributed to natural objects is the stone known as amethyst. The ancient Indian name of this stone had the sound represented by its present name. In Greek this sound happens to mean "not intoxicated"; hence, without more ado, the ancients declared that the amethyst was a preventive of, and a cure for, drunkenness.

ELEPHANTS

In the novel by that clever but contradictious writer, Sam Butler, entitled "The Way of All Flesh," an amiable and philosophically minded old gentleman, who pervades the story, states that when one feels worried or depressed by the incidents of one's daily life, great comfort may be derived from an hour spent at the Zoological Gardens in company with the larger mammalia. He ascribes to them a remarkable soothing influence, and I am inclined to agree with him. I am not prepared to decide whether the effect is due to the example of patience under adversity offered by these animals, or whether it is perhaps their tranquil indifference to everything but food, coupled with their magnificent success in attaining to such dignity of size, which imposes upon me and fills me for a brief space with resignation and a child-like acquiescence in things as they are. The elephant stands first as a soothing influence, and then the giraffe, the latter having special powers, due to its beautiful eyes and agreeable perfume. Sometimes the hippopotamus may diffuse a charm of his own, an aura of rotund obesity, especially when he is bathing or sleeping; but there are moments when one has to flee from his presence. I never could get on very well with rhinoceroses, but the large deer, bison, and wild cattle have the quality detected by Mr. Butler. So has the gorgeous, well-grown tiger, in full measure, when he purrs in answer to one's voice: but the lion is pompous, irritable, and easily upset. He never purrs. He is unpleasantly and obscurely spotted. He seems to be afraid of losing his dignity, and to be conscious of the fact that his reputation—like that of some English officials—depends on the overpowering wig which he now wears, though his Macedonian forerunner had no such growth to give an illusive appearance of size and capacity to his head. However opinions may differ about these things, we will agree that the elephant (or "Oliphant," as he was called in France 400 years ago) is the most imposing, fascinating, and astonishing of all animals.

At the present day there are two species only of elephant existing on the earth's surface. These are the Indian (Fig. 6) (called Elephas indicus, but sometimes called Elephas maximus on account of the priority which belongs to that designation, although the Indian elephant is smaller than the other), and the African (Fig. 7) (called Elephas Africanus). In the wild state their area of occupation has become greatly diminished within historic times. The Indian elephant was hunted in Mesopotamia in the twelfth century b.c., and Egyptian drawings of the eighteenth dynasty show elephants of this species brought as tribute by Syrian vassals. To-day the Indian elephant is confined to certain forests of Hindoostan, Ceylon, Burma, and Siam. The African elephant extended 100 years ago all over South Africa, and in the days of the Carthaginians was found near the Mediterranean shore, whilst in prehistoric (late Pleistocene) times it existed in the south of Spain and in Sicily. Now it is confined to the more central and equatorial zone of Africa, and is yearly receding before the incursions and destructive attacks of civilised man.

At no great distance of time before the historic period, earlier, indeed, than the times of the herdsmen who used polished stone implements and raised great stone circles, namely, in the late Pleistocene period, we find that there existed all over Europe and North Asia and the northern part of America another elephant very closely allied to the Indian elephant, but having a bow-like outward curvature of the tusks, their points finally directed towards one another, and a thick growth of coarse hair all over the body. This is "the mammoth," the remains of which are found in every river valley in England, France and Germany, and of which whole carcases are frequently discovered in Northern Siberia, preserved from decay in the frozen river gravels and "silt." The ancient cave-men of France used the fresh tusks of the mammoth killed on the spot for their carvings and engravings, and from their time to this the ivory of the mammoth has been, and remains, in constant use. It is estimated that during the last two centuries at least 100 pairs of mammoths' tusks have been each year exported from the frozen lands of Siberia. In early mediæval times the trade existed, and some ivory carvings and drinking horns of that age appear to be fashioned from this more ancient ivory.

Already, then, within the human period we find elephants closely similar to those of our own time, far more numerous and more widely distributed than in our own day, and happily established all over the temperate regions of the earth—even in our Thames Valley and in the forests where London now spreads its smoky brickwork. When we go further back in time—as the diggings and surveying of modern man enable us to do—we find other elephants of many different species, some differing greatly from the three species I have mentioned, and leading us back by gradual steps to a comparatively small animal, about the size of a donkey, without the wonderful trunk or the immense tusks of the later elephants. By the discovery and study of these earlier forms we have within the last ten years arrived at a knowledge of the steps by which the elephant acquired in the course of long ages (millions of years) his "proboscis" (as the Greeks first called it), and I will later sketch that history.

But now let us first of all note some of the peculiarities of living elephants and the points by which the two kinds differ from one another. The most striking fact about the elephant is its enormous size. It is only exceeded among living animals by whales; it is far larger than the biggest bull, or rhinoceros, or hippopotamus. A fair-sized Indian elephant weighs two to three tons (Jumbo, one of the African species, weighed five), and requires as food 60 lb. of oats, 1-1/2 truss of hay, 1-1/2 truss of corn a day, costing together in this country about 5s.; whereas a large cart-horse weighs 15 cwt., and requires weekly three trusses of hay and 80 lb. of oats, costing together 12s. or about 1s. 8-1/2d. a day. It is this which has proved fatal to the elephant since man took charge of the world. The elephant requires so much food and takes so many years in growing up (twenty or more before he is old enough to be put to work), that it is only in countries where there is a super-abundance of forest in which he can be allowed to grow to maturity at his own "charges" (so to speak) that it is worth while to attempt to domesticate and make use of him. For most purposes three horses are more "handy" than one elephant. The elephant is caught when he is already grown up, and then trained. It is as a matter of economy that he is not bred in confinement, and not because there is any insuperable difficulty in the matter. Occasionally elephants have bred in menageries.

There is no doubt that the African elephant at the present day grows to a larger size than the Indian, though it was the opinion of the Romans of the Empire that the Indian elephant was the more powerful, courageous, and intelligent of the two. It seems next to impossible to acquire at the present day either specimens or trustworthy records of the largest Indian elephants. About 10 ft. 6 ins. at the shoulder seems to be the maximum, though they are dressed up by their native owners with platforms and coverings to make them look bigger. In India the skin of domesticated individuals is polished and carefully stained, like an old boot, by the assiduity of their guardians, so that a museum specimen of exceptional size, fit for exhibition and study, cannot be obtained. On the other hand, the African elephant not unfrequently exceeds a height of 11 ft. at the shoulder. With some trouble I obtained one exceeding this measurement direct from East Africa for the Natural History Museum, where it now stands. It seems highly probable that this species occasionally exceeds 12 ft. in height. On the ground, between the great African elephant's fore and hind legs, in the museum, I placed a stuffed specimen of the smallest terrestrial mammal—the pigmy shrew-mouse. It is worth while thus calling to mind that the little animal has practically every separate bone, muscle, blood-vessel, nerve, and other structure present in the huge monster compared with it—is, in fact, built closely upon the same plan, and yet is so much smaller that it is impossible to measure one by the other. The mouse is only about one fifth the length of the elephant's eye. According to ancient Oriental fable, the mouse and the dragon were the only two animals of which the elephant was afraid.

The African elephant has much larger tusks relatively to his size than the Indian, and both males and females have them, whereas the Indian female has none. A very fine Indian elephant's tusk weighs from 75 lb. to 80 lb. The record for an African elephant's tusk was (according to standard books) 180 lb. But I obtained ten years ago for the museum, where it now may be seen, an African elephant's tusk weighing 228-1/2 lb. Its fellow weighed a couple of pounds less. It measures 10 ft. 2 in. in length along the curvature. This tusk was recognised by Sir Henry Stanley's companion, Mr. Jephson, when he was with me in the museum, as actually one which he had last seen in the centre of Africa. He told me that he had, in fact, weighed and measured this tusk in the treasury of Emin Pasha, in Central Africa, when he went with Stanley to bring Emin down to the coast. As will be remembered, Emin had no wish to go to the coast, but returned to his province. He was subsequently attacked and murdered by an Arab chief, who appropriated his store of ivory, and in the course of time had it conveyed to the ivory market at Zanzibar. The date of the purchase there of the museum specimen corresponds with the history given by Mr. Jephson.

The African elephant (as could be seen by comparing the small one living in Regent's Park with its neighbours) has a sloping forehead graduating into the trunk or proboscis, instead of the broad, upright brow of the Indian. He also has very much larger ears, which lie against the shoulders (except when he is greatly excited) like a short cape or cloak (see Fig. 7). These great ears differ somewhat in shape in the elephants of different parts of Africa, and local races can be distinguished by the longer or shorter angle into which the flap is drawn out. The grinding teeth of the two elephants differ very markedly, but one must see these in a museum. The grinders are very large and long (from behind forwards), coming into place one after the other. Each grinder occupies, when fully in position, the greater part of one side of the upper or of the lower jaw. They are crossed from right to left by ridges of enamel, like a series of mountains and valleys, which gradually wear down by rubbing against those of the tooth above or below. The biggest grinder of the Indian elephant has twenty-four of these transverse ridges, whilst that of the African has only eleven, which are therefore wider apart (see Fig. 8). An extinct kind of elephant—the mastodon—had only five such ridges on its biggest grinders, and four or only three on the others. Other ancestral elephants had quite ordinary-looking grinders, with only two or three irregular ridges or broad tubercles. Both the Indian and African elephant have hairless, rough, very hard, wrinkled skins. But the new-born young are covered with hair, and some Indian elephants living in cold, mountainous regions appear to retain a certain amount of hair through life. The mammoth (which agreed with the Indian elephant in the number of ridges on its grinders and in other points) lived in quite cold, sub-Arctic conditions, at a time when glaciers completely covered Scandinavia and the north of our islands as well as most of Germany. It retained a complete coat of coarse hair throughout life. The young of our surviving elephants only exhibit transitorily the family tendency.

The last mammoth probably disappeared from the area which is now Great Britain about 150,000 years ago. It might be supposed that no elephant was seen in England again until the creation of "menageries" and "zoological gardens" within the last two or three hundred years. This, however, is by no means the case. The Italians in the middle ages, and through them the French and the rulers of Central Europe, kept menageries and received as presents, or in connection with their trade with the East and their relations with Eastern rulers, frequent specimens of strange beasts from distant lands. Our King Henry I, had a menagerie at Woodstock, where he kept a porcupine, lions, leopards, and a camel! The Emperor Charlemagne received in 803 a.d. from Haroun al Raschid, the Caliph of Bagdad, an elephant named Abulabaz. It was brought to Aix-la-Chapelle by Isaac the Jew, and died suddenly in 810. Some four and a half centuries later (in 1257), Louis IX, of France, returning from the Holy Land, sent as a special and magnificent present to Henry III, King of England (according to the chronicle of Matthew Paris), an elephant which was exhibited at the Tower of London. It was supposed by the chronicler to be the first ever brought to England, and indeed the first to be taken beyond Italy, for he did not know of Charlemagne's specimen. In 1591 King Henry IV of France, wishing to be very polite to Queen Elizabeth of England, and apparently rather troubled by the expense of keeping the beast himself, sent to her, having heard that she would like to have it, an elephant which had been brought from the "Indies" and landed at Dieppe. He declared it to be the first which had ever come into France, but presented it to Her Majesty "as I would most willingly present anything more excellent did I possess it." Thenceforward elephants were from time to time exhibited at the Tower, together with lions and other strange beasts acquired by the Crown.

None of these elephants were, however, "the first who ever burst" into remote Britain after the mammoths had disappeared, and we were separated from Europe by the geological changes which gave us the English Channel—La Manche. Though Julius Cæsar himself does not mention it, it is definitely stated by a writer on strategy named Polyænus, a friend of the Emperor Marcus Aurelius, but not, I am sorry to say, an authority to whose statements historians attach any serious value—that Cæsar made use of an elephant armed with iron plates and carrying on its back a tower full of armed men to terrify the ancient Britons when he crossed the Thames—an operation which he carried out, I believe, somewhere between Molesey and Staines.

Elephants are often spoken of as "Ungulates," and classed by naturalists with the hoofed animals (the odd toed tapirs, rhinoceroses, and horses, and the even-toed pigs, camel, cattle, and deer). But there is not much to say in defence of such an association. The elephants have, as a matter of fact, not got hoofs, and they have five toes on each foot. The five toes of the front foot have each a nail, whilst usually only four toes of the hind foot have nails. A speciality of the elephant is the great circular pad of thick skin overlying fat and fibrous tissue, which forms the sole of the foot and bears the animal's enormous weight. This buffer-like development of the foot existed in some great extinct mammals (the Dinoceras family, of North America), but is altogether different from the support given by a horse's hoof or the paired shoe-like hoofs of great cattle or the three rather elegant hoofed toes of the rhinoceros.

The Indian elephant likes good, solid ground to walk on, and when he finds himself in a boggy place will seize any large objects (preferably big branches of trees) and throw them under his feet to prevent himself sinking in. Occasionally he will remove the stranger who is riding on his back and make use of him in this way. The circumference of the African elephant's fore-foot is found by hunters to be half the animal's height at the shoulder, and is regarded as furnishing a trustworthy indication of his stature.

The legs of the elephant differ from those of more familiar large animals in the fact that the ankle and the wrist (the so-called knee of the horse's foreleg) are not far above the sole of the foot (resembling man's joints in this respect), whilst the true knee-joint (called "the stifle" in horses)—instead of being, as in horses, high up, close against the body, strongly flexed even when at rest, and obscured by the skin—is far below the body, free and obvious enough. In fact, the elephant keeps the thigh and the upper arm perpendicular and in line with the lower segment of the limb when he is standing, so that the legs are pillar-like. But he bends the joints amply when in quick movement. The hind legs seen in action resemble, in the proportions of thigh, foreleg, and foot, and the bending at the knee and ankle, very closely those of a man walking on "all fours." The elephant as known in Europe more than 300 years ago was rarely seen in free movement. He was kept chained up in his stall, resting on his straight, pillar-like legs and their pad-like feet. And with that curious avidity for the marvellous which characterized serious writers in those days to the exclusion of any desire or attempt to ascertain the truth, it was coolly asserted, and then commonly believed, that the elephant could not bend his legs. Shakespeare—who, of course, is merely using a common belief of his time as a chance illustration of human character—makes Ulysses say (referring to his own stiffness of carriage) ("Troilus and Cressida," Act II) "The elephant hath joints, but none for courtesy; his legs are legs for necessity, not for flexure." An old writer says: "The elephant hath no joints, and, being unable to lye down, it lieth against a tree, which, the hunters observing, do saw almost asunder; whereon the beast relying—by the fall of the tree falls also down itself, and is able to rise no more." Another old writer (Bartholomew, 1485), says, more correctly: "When the elephant sitteth he bendeth his feet; he bendeth the hinder legs right as a man."

A writer of 120 years later in date (Topsell) says: "In the River Ganges there are blue worms of sixty cubits long having two arms; these when the elephants come to drink in that river take their trunks in their hands and pull them off. At the sight of a beautiful woman elephants leave off all rage and grow meek and gentle. In Africa there are certain springs of water which, if at any time they dry up, they are opened and recovered again by the teeth of elephants." The blue worm of the Ganges referred to is no doubt the crocodile; both in India and Africa animals coming to the rivers to drink are seized by lurking crocodiles, who fix their powerful jaws on to the face (snout or muzzle) of the drinking animal and drag it under the water. Thus the fable has arisen of the origin of the elephant's trunk as recounted by Mr. Rudyard Kipling. A young elephant (before the days of trunks), according to this authority, when drinking at a riverside had his moderate and well-shaped snout seized by a crocodile. The little elephant pulled and the crocodile pulled, and by the help of a friendly python the elephant got the best of it. He extricated himself from the jaws of death. But, oh! what a difference in his appearance! His snout was drawn out so as to form that wonderful elongated thing with two nostrils at the end which we call the elephant's trunk, and was henceforth transmitted (a first-rate example of an "acquired character") to future generations! The real origin of the elephant's trunk is (as I will explain later) a different one from that handed down to us in the delightful jungle-book. I do not believe in the hereditary transmission of acquired modifications!

Topsell may or may not be right as to the result produced on elephants by the sight of a beautiful woman. In Africa the experiment would be a difficult one, and even in India inconclusive. Topsell seems, however, to have come across correct information about the digging for water by an African elephant by the use of his great tusks—those tusks for the gain of which he is now being rapidly exterminated by man. Serious drought is frequent in Africa, and a cause of death to thousands of animals. African elephants, working in company, are known to have excavated holes in dried-up river beds to the depth of 25 feet in a single night in search of water. It is probable that the Indian elephant's tusk would not be of service in such digging, and it is to be noted that he is rather an inhabitant of high ground and table-lands than of tropical plains liable to flood and to drought. The tusk of the Indian elephant has become merely a weapon of attack for the male, and there are even local breeds in which it is absent in the males as well as in the females. The mammoth was a near cousin of the Indian elephant, and inhabited cold uplands and the fringes of sub-Arctic forests, on which he fed. His tusks were very large, and curved first outward and then inward at the tips. They would not have served for heavy digging, and probably were used for forcing a way through the forest and as a protection to the face and trunk.

The trunk of the elephant was called "a hand" by old writers, and it seems to have acted in the development of the elephant's intelligence in the same way as man's hand has in regard to his mental growth, though in a less degree. The Indian elephant has a single tactile and grasping projection (sometimes called "a finger") placed above between the two nostrils at the end of the trunk; the African elephant has one above and one below. I have seen the elephant pick up with this wonderful trunk with equal facility a heavy man and then a threepenny piece.

The intelligence of the elephant is sometimes exaggerated by reports and stories; sometimes it is not sufficiently appreciated. It is not fair to compare the intelligence of the elephant with that of the dog—bred and trained by man for thousands of years. So far as one can judge, there is no wild animal, excepting the higher apes, which exhibits so much and such varied intelligence as the elephant. It appears that from early tertiary times (late Eocene) the ancestors of elephants have had large brains, whilst, when we go back so far as this, the ancestors of nearly all other animals had brains a quarter of the size (and even less in proportion to body-size) which their modern representatives have. Probably the early possession of a large brain at a geological period when brains were as a rule small is what has enabled the elephants not only to survive until to-day, but to spread over the whole world (except Australia), and to develop an immense variety and number of individuals throughout the tertiary series in spite of their ungainly size. It is only the yet bigger brain of man which (would it were not so!) is now at last driving this lovable giant, this vast compound of sagacity and strength, out of existence. The elephant—like man standing on his hind legs—has a wide survey of things around him owing to his height. He can take time to allow of cerebral intervention in his actions since he is so large that he has little cause to be afraid and to hurry. He has a fine and delicate exploring organ in his trunk, with its hand-like termination; with this he can, and does, experiment and builds up his individual knowledge and experience. Elephants act together in the wild state, aiding one another to uproot trees too large for one to deal with alone. They readily understand and accept the guidance of man, and with very small persuasion and teaching execute very dextrous work—such as the piling of timber. If man had selected the more intelligent elephants for breeding over a space of a couple of thousand years a prodigy of animal intelligence would have resulted. But man has never "bred" the elephant at all.

The Greeks and Romans knew ivory first, and then became acquainted with the elephant. The island of Elephantina in the Nile was from the earliest times a seat of trade in the ivory tusks of the African elephant, and so acquired its name. Herodotus is the first to mention the elephant itself; Homer only refers to the ivory by the word "elephas." Aristotle in this, as in other matters, is more correct than later writers. He probably received first-hand information about the elephant from Alexander and some of his men after their Indian expedition. The Romans had an unpleasant first personal experience of elephants when Pyrrhus, King of Epirus, landed a number with his army and put the Roman soldiers to flight. But the Romans then, and continually in after-times, showed their cool heads and sound judgment in a certain contempt for elephants as engines of war. They soon learned to dig pits on the battlefield to entrap the great beasts, and they deliberately made for the elephants' trunks, hewing them through with their swords, so that the agonised and maddened creatures turned round and trampled down the troops of their own side. The Romans only used them subsequently to terrify barbaric people, and as features in military processions. But Eastern nations used them extensively in war. In a.d. 217 Antiochus the Great brought 217 elephants in his army against 73 employed by Ptolemy, at what was called "the Battle of the Elephants." The battle commenced by the charging head to head of the opposing elephants and the discharge of arrows, spears and stones by the men in the towers on their backs.

An interesting question has been raised as to whether the elephants used by the Carthaginians were the African species or the Indian. There is no doubt that the elephants of Pyrrhus and those known to Alexander were the Indian, though they were taken in those days much to the West of India, namely, in Mesopotamia, and it would not have been difficult for the Carthaginians to convey Indian elephants, which had certainly been brought as far as Egypt, along the Mediterranean coast. An unfounded prejudice as to the want of docility of the African elephant has favoured the notion that the Carthaginians used the Indian elephant. As a matter of fact, no one in modern times has tried to train the African elephant, except here and there in a zoological garden. Probably the Indian "mahout," or elephant trainer could, if he were put to it, do as much with an African as he does with an Indian elephant. It would be an interesting experiment. In the next place, there is decisive evidence that it was the African elephant which the

Carthaginians used, since we have a Carthaginian coin (Fig. 7) on which is beautifully represented—in unmistakable modelling—the African elephant, with his large triangular cape-like ears and his sloping forehead. In the time of Hannibal there were stables for over 300 of these elephants at Carthage, and he took fifty with him to the South of France with his army for the Italian invasion. He only got thirty-seven safely over the Rhone, and all but a dozen or so died in the terrible passage of the Alps. After the battle of Trebia he had only eight left, and when he had crossed the Apennines there was only one still alive. On this Hannibal himself rode.

Since the period when the white chalk which now forms our cliffs and hills was deposited at the bottom of a vast and deep ocean—the sea bottom has been raised, the chalk has emerged and risen on the top of hills to 800 feet in height in our own islands, and to ten times that height elsewhere, and during that process sands and clays and shelly gravels have been deposited to the thickness of some 2,800 feet by seas and estuaries and lakes, which have come and gone on the face of Europe and of other parts of the world as it has slowly sunk and slowly risen again. The last 200 feet or so of deposits we call the Pleistocene or Quaternary; the rest are known as the Tertiary strata. They are only a small part of the total thickness of aqueous deposit of stratified rock—which amounts to 60,000 feet more before the earliest remains of life in the Cambrian beds are reached, whilst older than, and therefore below this, we have another 50,000 feet of water-made rock which yields no fossils—no remains of living things, though living things were certainly there! Our little layer of Tertiary strata on the top is, however, very important. It took several million years in forming, although it is only one-fortieth of the whole thickness of aqueous deposit on the crust of the earth. We divide it into Pliocene, Miocene, and Eocene, and each of these into upper, middle, and lower, the Eocene being the oldest. Our London clay and Woolwich sands are lower Eocene; there is a good deal of Miocene in Switzerland and Germany, whilst the Pliocene is represented by whole provinces of Italy, parts of central France, and by the White and Red "crags" of Suffolk.[5]

It is during this Tertiary period that the mammals—the warm-blooded, hairy quadrupeds, which suckle their young—have developed (they had come into existence a good deal earlier), and we find the remains of ancestral forms of the living kinds of cattle, pigs, horses, rhinoceroses, tapirs, elephants, lions, wolves, bears, etc., embedded in the successive layers of Tertiary deposits. Naturally enough, those most like the present animals are found in late Pliocene, and those which are close to the common ancestors of many of the later kinds are found in the Eocene, whilst we also find, at various levels of the Tertiary deposit, remains of side-branches of the mammalian pedigree, which, though including very powerful and remarkable beasts, have left no line of descent to represent them at the present day. We have been able to trace the great modern one-toed horses, zebras, and asses, with their complicated pattern of grinding-teeth back by quite gradual steps (represented by the bones and teeth of fossil kinds of horses), to smaller three-toed animals with simpler tuberculated teeth, and even, without any marked break in the series, to a small Eocene animal (not bigger than a spaniel) with four equal-sized toes on its front foot, and three on its hind foot. We know, too, a less direct series of intermediate forms leading beyond this to an animal with five toes on each foot and "typical" teeth. In fact, no one doubts that (leaving aside a few difficult and doubtful cases) all such big existing mammals, as I mentioned above, as well as monkeys and man, are derived from small mammals—intermediate in most ways between a hedgehog and a pig—which flourished in very early Eocene times, and had five toes on each foot, and "a typical dentition." Even the elephants came from such a small ancestral form. The common notion that the extinct forerunners of existing animals were much bigger than recent kinds, and even gigantic, is not in accordance with fact. Some extinct animals were of very great size—especially the great reptiles of the period long before the Tertiaries, and before the chalk. But the recent horse, the recent elephant, the giraffe, the lions, bears, and others, are bigger—some much bigger—than the ancestral forms, to which we can trace them by the wonderfully preserved and wonderfully collected and worked-out fossilised bones discovered in the successive layers of the Pliocene, Miocene, and Eocene strata, leading us as we descend to more primitive, simplified, and smaller ancestors.

It is easy to understand the initial character of the foot of the early ancestral mammals. It had five toes. By the suppression or atrophy of first the innermost toe, then of the outermost, you find that mammals may first acquire four toes only, and then only three, and by repeating the process the toes may be reduced to two, or right away to one, the original middle toe. There is no special difficulty about tracing back the elephants in so far as this matter is concerned, since they have kept (like man and some other mammals) the full typical complement of five toes on each foot.

But I must explain a little more at length what was the "typical dentition,"—that is to say, the exact number and form of the teeth in each half of the upper and the lower jaw of the early mammalian ancestor of lower Eocene times, or just before. The jaws were drawn out into a snout or muzzle, an elongated, protruding "face," as in a dog or deer or hedgehog, and there were numerous teeth set in a row along the gums of the upper and the lower jaw. The teeth were the same in number, in upper and in lower jaw, and so formed as to work together, those of the lower jaw shutting as a rule just a little in front of the corresponding teeth of the upper jaw. There were above and below, in front, six small chisel-like teeth, which we call "the incisors." At the corner of the mouth above and below on each side flanking these was a corner tooth, or dog-tooth, a little bigger than the incisors, and more pointed and projecting. These we call "the canines," four in all. Then we turn the corner of the mouth-front, as it were, and come to the "grinders," cheek-teeth or molars. These are placed in a row along each half of upper and lower jaw. In our early mammalian ancestor they were seven in number, with broader crowns than the peg-like incisors and canines, the bright polished enamel of the crown being raised up into two, three or four cone-like prominences. The back grinders are broader and bigger than those nearer the dog-tooth. The three hindermost grinders in each half of each jaw are not replaced by "second" teeth, whilst all the other teeth are.

Now this typical set of teeth—consisting of twenty-eight grinders, four canines, and twelve incisors—is not found complete in many mammals at the present day, though it is found more frequently as we go back to earlier strata.[6] Though some mammals have kept close to the original number, they have developed peculiar shape and qualities in some of the teeth as well as changes in size. The common pig still keeps the typical number (Fig. 10), but he has developed the corner teeth or canines into enormous tusks both in the upper and lower jaw, and the more anterior grinders have become quite minute. The cats (lions and tigers included) have kept the full number of incisors (see Figs. 21 and 22, pp. 103, 104); they have developed the four canines into enormous and deadly stabbing "fangs," and they have lost all the grinders but three in each half of the lower jaw and four in each half of the upper jaw (twelve instead of twenty-eight), and these have become sharp-edged so as to be scissor-like in their action, instead of crushing or grinding. Man and the old-world monkeys have lost an incisor in each half of each jaw (see Pls. VI and VII); they retain the canines, but have only five molars in each half of each jaw (twenty in all instead of twenty-eight). Most of the mammals—whatever change of number and shape has befallen their teeth in adaptation to their different requirements as to the kind of food and mode of getting it—have retained a good long pair of jaws and a snout or muzzle consisting of nose, upper jaw, and lower jaw, projecting well in front of the eyes and brain-case. Man is remarkable as an exception. In the higher races of men the jaws are shorter than in the lower races, and project but very little beyond the vertical plane of the eyes, whilst the nose projects beyond the lips. Another exception is the elephant. This is most obvious when the prepared bony skull and lower jaw are examined, but can be sufficiently clearly seen in the living animal. The lower jaw and the part of the upper jaw against which it and its grinders play is extraordinarily short and small. The elephant has, in fact, no projecting bony jaw at all, no bony snout, its chin does not project more than that of an old man, and even the part of the upper jaw into which its great tusks are set does not bend forward far from the perpendicular (Fig. 9).

The elephant (see Fig. 9) has no sign of the six little front teeth (incisors) above and below which we find in the typical dentition and in many living mammals, nor of the corner teeth (dog-teeth, or canines). In the upper jaw in front there is the one huge tusk on each side, and in the lower jaw no front teeth at all! Then as to the grinders. In the elephant these are enormous, with many transverse ridges on the elongated crown, and so big that there is only room for one at a time in each half of upper and lower jaw. Six of these succeed one another in each half of each jaw, and correspond (though greatly altered) to six of the seven grinders of the typical dentition. Are there amongst older fossil elephants and animals like elephants any which have an intermediate condition of the teeth, connecting the extremely peculiar teeth of the modern elephants with the typical dentition such as is approached by the pig, the dog, the tapir, and the hedgehog? There are such links. We know a great many elephants from Pleistocene and Pliocene strata—some from European localities, more from India, and some from America. A little elephant not more than 3 feet high when adult is found fossil in the island of Malta; other species were a little larger than the living African elephant. Whilst the Indian elephant has as many as twenty-four cross-ridges on its biggest grinding tooth (Fig. 8) there is a fossil kind which has only six such ridges. But besides true elephants we know from the Pliocene, Miocene, and Upper Eocene of the old world, the remains of elephant-like creatures (some as big as true elephants), which are distinguished by the name "Mastodon" (Fig. 11). And, in fact, we are conducted through a series of changes of form by ancient elephant-like creatures which are of older and older date as we pass along the series, and are known as (1) Mastodon, (2) Tetrabelodon, (3) Palæomastodon, (4) Meritherium, until we come to something approaching the general form of skull and skeleton and the typical dentition of the early mammalian ancestor. Mastodons of several species are found in Pliocene strata in Europe and Asia; detached teeth are found in Suffolk. One species actually survived (why, we do not know) in North America into the early human period, and whole skeletons of it are dug out from the morasses such as that of "Big-bone Lick." The Mastodons had a longer jaw and face than the elephants, though closely allied to them. They bring one nearer to ordinary mammals in that fact, and also in having (when young) two front teeth or incisors in the lower jaw. Their grinders had the crowns less elongated than those of the elephants, and there were only five cross-ridges—on the biggest—and these ridges tend to divide into separate cones (Fig. 8). So here, too, we are approaching the ordinary mammals, of which we may keep the pig and the tapir in mind as samples. But the Mastodons still had the great trunk and huge tusks of the elephants.

Next we must look at Tetrabelodon (Fig. 12), and it is this creature which has really revealed the history of the strange metamorphosis by which elephants were produced. The Tetrabelodon is known as "the long-jawed mastodon," because, as was shown in a wonderfully well-preserved skeleton from the lower Pliocene of the centre of France, set up in the Paris Museum, it had a lower jaw of enormous length, ending in two large horizontally directed teeth (Fig. 12). Instead of a lower jaw a foot long, as in an elephant or in the common kind of mastodon—this long-jawed kind had a lower jaw 5 feet or 6 feet long! The tusks of the upper jaw were large, and nearly horizontal in direction, bent downwards a little on each side of the long lower jaw. This lower jaw seemed incomprehensible, almost a monstrosity—until it occurred to me that it exactly corresponds to the elongated upper lip and nose which we call the elephant's trunk—and that the trunk of "Tetrabelodon" must have rested on his long lower jaw. In descending to Tetrabelodon we leave behind us the elephants with hanging unsupported trunk; the lower jaw here is of sufficient length to support the great trunk. When the lower jaw shortened in the later mastodons and elephants the trunk did not shorten too, but remained free and depending, capable of large movement and of grasping with its extremity. Photographs, casts, and actual specimens of the extraordinary skull of the long-jawed mastodon or Tetrabelodon and of the creatures mentioned below may be seen in the Natural History Museum.

Lastly we have the wonderful series of discoveries made about twelve years ago by Dr. Andrews (of the Natural History Museum) of elephant-like creatures in the upper Eocene of the Fayoum Desert of Egypt. Palæomastodon (the name given by Dr. Andrews to one of them) is a "pig-like" mastodon, with an elongated, bony face, the tusks of moderate size, and the lower jaw not projecting more than a few inches beyond them, so that the proboscis is quite short and rests well on it (Fig. 13). This animal had six moderate sized grinders (molars or cheek-teeth) on each side of each jaw in position simultaneously, as may be seen in the complete skull shown in Fig. 14. Of other teeth it had only the two moderate-sized front tusks above and two very big, chisel-like "incisors" in the front of the lower jaw. Exactly how these were used and for what food no one has yet made out.

The remains, which finally bring the elephants into line with the ordinary mammals with typical dentition, were discovered also by Dr. Andrews and named "Meritherium" by him, signifying "the beast of the Lake Meris." This creature is not bigger than a tapir, and had the shape of head and face which we see in that and the ordinary hoofed animals (Fig. 15). It had no trunk, and whilst it had six small and simplified mastodon-like grinders in each half of each jaw, it had six incisors in the upper jaw and a canine or corner tooth on each side. In the lower jaw there were only two large incisors besides the cheek-teeth or grinders. Not the least interesting point about Meritherium is that it tells us which of the front upper teeth have become the huge tusks of the later elephants. Counting from the middle line there are in Meritherium three incisors right and three left. The second of these upper teeth on each side is much larger than the others. It is this (seen in Fig. 15) which has grown larger and larger in later descendants of this primitive form and become the elephant's tusk, whilst all the others have disappeared.

We now know the complete series of steps connecting elephants with ordinary trunkless, tuskless mammals. The transition from the "beast of Meris" on the one hand to the common typidentate mammalian ancestor, and on the other hand to the elephants, is easy, and requires no effort of the imagination. His short muzzle (upper and lower jaw), first elongated step by step to a considerable length, giving us Palæomastodon (Fig. 13). Then the lower jaw shrunk and became shorter than it was at the start, and the rest of the muzzle (the front part of the upper jaw, carrying with it the nostrils), drooped and became the mobile muscular elephant's trunk!

[5] I am inclined to think that the line between Pliocene and Pleistocene or Quaternary ought, in this country, to be drawn between the White and Red Crag of Suffolk. Glacial conditions set in and were recurrent from the commencement of the Red Crag deposit onwards.

[6] Mammals having the number and form of teeth which I have just described as typical—or such modification of it as can easily be produced by suppression of some teeth and enlargement of others—are called Typidentata. On the other hand, the whales, the sloths, ant-eaters, and armadilloes, as also the Marsupials, are called Variodentata, because we cannot derive their teeth from those of the Typidentate ancestor. They form lines of descent which separated from the other mammals before the Typidentate ancestor of all, except the groups just named, was evolved.

A STRANGE EXTINCT BEAST

The terraces of gravel deposited by existing rivers and the deposits in caverns in the limestone regions of Western Europe—the so-called "Pleistocene" strata—contain, besides the flint weapons of man and rare specimens of his bones, the remains of animals which are either identical with those living at the present day (though many of them are not living now in Europe) or of animals very closely similar to living species. Thus we find the bones of horses like the wild horse of Mongolia, of the great bull (the Urus of Cæsar), of the bison, of deer and goats, of the Siberian big-nosed antelope, of the musk-ox (now living within the Arctic circle), of the wild boar, of the hippopotamus (like that of the Nile), and of lions, hyenas, bears, and wolves. The most noteworthy of the animals like to, but not identical with, any living species are the mammoth, which is very close to the Indian elephant, but has a hairy coat; the hairy rhinoceros, like, but not quite the same as, the African square-mouthed rhinoceros; and the great Irish deer, which is like a giant fallow-deer. These three animals are really extinct kinds or species, but are not very far from living kinds. In fact, the most recent geological deposits do not contain any animals so peculiar, when compared with living animals, as to necessitate a wide separation of the fossil animal from living "congeners" by the naturalist who classifies animals and tries to exhibit their degrees of likeness and relationship to one another by the names he adopts for them. The mammoth is a distinct "species" of elephant. It requires, it is true, a "specific" or "second" name of its own; but it belongs to the genus elephant. Hence we call it Elephas primigenius, whilst the living Indian elephant is Elephas Indicus. The reader is referred to the preceding chapter for further notes about elephants.

The strata next below the Pleistocene gravels and cave deposits are ascribed to the "Pliocene age"—older than these are the "Miocene" and the

"Eocene," and then you come to the Chalk, a good white landmark separating newer from older strata.

We know now in great detail the skeletons and jaws of some hundreds of kinds of extinct animals of very different groups found in the Eocene, the Miocene, the Pliocene, and the Pleistocene layers of clays, sands, and gravels of this part of the world. Nothing very strange or unlike what is now living is found in the Pleistocene—the latest deposits—but when we go further back strange creatures are discovered, becoming stranger and less like living things as we pass through Pliocene to Miocene, and on—downwards in layers, backwards in time—to the Eocene.

Though the past history of the Mediterranean sea shows that it was formerly not so extensive as it is now, and that there were junctions between Europe and Africa across its waters, yet the deeper parts of that sea are very ancient, and some of the islands have long been isolated. In Malta the remains of extraordinary species of minute elephants have been found, one no larger than a small donkey, and in the island of Cyprus an English lady, Miss Dorothea Bate, has discovered the bones of a pigmy hippopotamus (like that still living in Liberia) no larger than a sheep. Miss Bate some three years ago heard of the existence of a bone-containing deposit of Pleistocene age in limestone caverns and fissures in the island of Majorca, and with the true enthusiasm of an explorer determined to carry on some "digging" there and see what might turn up. In the following spring she was there, and obtained a number of bones, jaws, and portions of skulls, which appeared at first sight to be those of a small goat. Its size may be gathered from the fact that its skull is six inches long. These and the bones of a few small finches were all that rewarded her pains. The bones of fossil goats (of living species) are found in caves at Gibraltar and in Spain; so at first the result seemed disappointing. But on carefully clearing out the specimens and examining them in London, Miss Bate found that the supposed goat bones obtained by her in Majorca were really those of a new and most extraordinary animal, to which (in a paper published in the "Geological Magazine" in September, 1910) she has given the name "Myotragus balearicus."

I must ask the reader now to look at the figures here given (Figs. 16 and 17) of the skull and the lower jaw of a goat. The lower jaw might (except for size) pass for that of a sheep, ox, antelope or deer. They are all alike. There are on each side six grinding cheek-teeth (molars), and then as we pass to the front we find a long toothless gap until we come to the middle line where the two halves of the jaw unite. There we see a little semicircular group of eight chisel-like teeth, which work against the toothless pad of the upper jaw opposed to them, and are the instruments by which these animals, with an upward jerk of the head, "crop" the grass and other herbage on which they feed, to be afterwards triturated by the grinding cheek teeth. A vast series of living and of fossil animals, called the Ruminants—including the giraffes, the antler-bearing forms called deer, the cavicorn or sheath-horned bovines, ovines and caprines, and the large series of antelopes of Africa and India—all have precisely this form of jaw, this number and shape and grouping of the teeth. Now let me call to mind the lower jaw of a hare or rabbit or rat (Figs. 18 and 19). There we find on each side the group of grinding cheek-teeth, with transverse ridges on their crowns, and a long, toothless gap before we arrive at the front teeth. But the front teeth are only two in number, one on each side, close to each other, very large, and each with a tremendously long, deeply set root. They meet a similar pair of teeth in the upper jaw, and give the hare, rabbit, rats, mice, beavers, and porcupines the power of "gnawing" tough substances. These animals are hence called Rodents, or gnawers, and the two great front teeth are called "rodent-teeth." No two arrangements of teeth could be much more unlike than are the group of eight little chisel-like teeth of the lower jaw of the Ruminants and the two enormous gnawing teeth of the Rodents. Apparently the two rodent incisors, or front teeth, of the lower jaw of the rat correspond to the two middle incisors of the Ruminant's lower jaw; the other front teeth of the Ruminant have atrophied, disappeared altogether. The rodent condition has been developed from that of an ancestor which had several front teeth and not two large ones only; but we have not at present found the intermediate steps.

The reader should compare the teeth of the goat and the large rat here pictured with the more typical and complete series of the pig, given in Fig. 10, p. 84. The pig's teeth are the same in number as those of the ancestral primitive typidentate mammal, and their form is near to that of the ancestor's teeth.

Now I come to the extraordinary interest of Miss Bate's goat-like or antelope-like animal from Majorca. Although it is shown by its skull (Fig. 20) and other bones to be distinctly one of the sheath-horned Ruminants, very like a small goat or antelope, the lower jaw, of which there are several specimens, does not present in front the little group of eight small chisel-like "cropping" teeth, but, instead, two enormous rodent teeth placed side by side, very deeply fixed in the jaw, and quite like those of some rat-like animals in shape. Hence the name given to this little marvel by Miss Bate— "Myotragus," "the rat-goat." This strange little animal also differs from goats and antelopes in having proportionately much thicker and shorter "feet" (cannon-bones) than they have.

If the remains of this strange little creature had turned up in more ancient strata—in Pliocene or Miocene—it would have not been quite so astonishing. But it would be still very remarkable, since it has all the characters of a goat-like creature in the shape of its skull, its bony horncores, its limb-bones, and its cheek-teeth; and yet, as it were monstrously and in a most disconcerting way, protrudes from its lower jaw two great rats' teeth. Nothing like it or approaching it or suggesting it, is known among recent or fossil Ruminants. They all without exception have a lower jaw with the teeth of the exact number and grouping which you may see in a sheep's lower jaw. We know hundreds of them, both living and fossil, many from the Pleistocene, others from Pliocene deposits, and even from the still older Miocene, but all keep to the one pattern of lower jaw and lower jaw teeth. It is only in this little island of Majorca, surrounded by very deep water and not known to have nurtured any other animal so large in size either in recent or geologic times, that we come upon a Ruminant with horns like a goat's, but with great rat-like front teeth in place of the semicircle of eight little cropping toothlets. The wonderful thing is that the bones found by Miss Bate are light and well preserved, evidently not very ancient—probably late Pleistocene in age.

The questions that arise are: Where did the rat-goat come from? How did this utterly peculiar change in a Ruminant's teeth come about? With regard to the second question, it is a matter of importance that although we have hitherto not discovered any Ruminants with this modification of the teeth, still less any cavicorn or sheath-horned Ruminant so altered, yet it is by no means rare amongst herbivorous mammals to find such rat-like teeth making their appearance, whilst the smaller side-teeth of the incisor group or front teeth disappear. The Australian kangaroos and wombats are a case in point—so is the lemur-like aye-aye of Madagascar (an insect eater). So is the Hyrax or "damian" of the Cape, and also the very ancient Plagiaulax from the præ-chalk Purbeck clay. But perhaps the best case for comparison with the ruminants is that of the rhinoceroses. There are a great many species and even genera of fossil and recent rhinoceroses. An old Miocene kind (called Hyracodon) has eight little teeth in the front of the lower jaw. In a Pliocene kind of rhinoceros (called R. incisivus) these are reduced to two, the middle two, which are of great size and project far forward—like those of the rat-goat of Majorca. Among living rhinoceroses the Indian species have these two front teeth, but smaller, whilst the square-mouthed African rhinoceros has none at all! This helps us, as a parallel, to understand "the strange case" of Myotragus. But, of course, the rhinoceroses are a distinct line of animal descent—remote from Ruminants. They are (like horses and tapirs) odd-toed hoofed beasts—not even-toed ones, as are pigs, camels, and ruminants.

On first considering the question of the origin of the rat-goat of Majorca, some naturalists will, no doubt, be tempted to suggest that it is a case of a sudden "sport," a "mutation" as they now call it, and not a result of gradual slowly developed reduction of the now lost teeth and correspondingly gradual enlargement of the two middle ones, taking many thousand generations to bring about. The fact that the rat-goat is found on an island cut off from competition with other animals will favour this view. On the other hand, there is the important and really remarkable fact that familiar as man has been for ages with Ruminants of many kinds—such as sheep, goats, cattle, deer—there is absolutely no case on record of an "oddity" or "monstrosity" resembling the rat-goat's condition occurring in the teeth of any of the hundreds of thousands of these animals killed and eaten by man, and therefore closely examined. Professor Bateson, who a few years ago ransacked the museums of Europe for instances of "discontinuous variation," or "sports," and wrote a valuable book on the subject, did not discover any example of the kind. Apart from the view, which is very generally held, that such sudden "mutations" as "rat-teeth in a ruminant" are—even should they occur—not perpetuated, we are not really in any way driven to suppose that the rat-goat of Majorca originated in that island. It is true that we know nothing like it in the Pliocene and Miocene of the Mediterranean region which could have been its immediate ancestor. But probably the ancestors of the rat-goat were slowly developed from a Miocene sheath-horned ruminant, a primitive sort of antelope in some part of North-west Africa, or in an extension of it now submerged in the Atlantic, and stragglers of this curious and now lost Ruminant stock were left in Majorca when in Miocene or early Pliocene times that island became detached from its Hispano-African connection.

VEGETARIANS AND THEIR TEETH

No mistake, said Huxley, is more frequently made by clever people than that of supposing that a cause or an opinion is unsound because the arguments put forward in its favour by its advocates are foolish or erroneous. Some of the arguments put forward in favour of the exclusive use by mankind of a vegetable diet can be shown to be based on misconception and error, and I propose now to mention one or two of these. But I wish to guard against the supposition that I am convinced in consequence that animal substances form the best possible diet for man, or that an exclusively vegetable diet may not, if properly selected, be advantageous for a large majority of mankind. That question, as well as the question of the advantage of a mixed diet of animal and vegetable substances, and the best proportion and quantity of the substances so mixed, must be settled, as also the question as to the harm or good in the habitual use of small quantities of alcohol, by definite careful experiment by competent physiologists, conducted on a scale large enough to give conclusive results. The cogency of the arguments in favour of vegetarianism which I am about to discuss is another matter.

In the first place it is very generally asserted by those who advocate a purely vegetable diet that man's teeth are of the shape and pattern which we find in fruit-eating or in root-eating animals allied to him. This is true. The warm-blooded hairy quadrupeds which suckle their young and are called "mammals" (for which word perhaps "beasts" is the nearest Anglo-Saxon equivalent) show in different groups and orders a great variety in their teeth. The birds of to-day have no teeth, the reptiles, amphibians, and fishes have usually simple conical or peg-like teeth, which are used simply for holding and tearing. In some cases the pointed pin-like teeth are broadened out so as to be button-like, and act as crushing organs for breaking up shell-fish.

The mammals alone have a great variety and elaboration of the teeth.

In shape and size, as well as in number, the teeth of mammals are very clearly related to the nature of their food in the first place, and secondly to their use as weapons of attack or of defence. When the surface of the cheek-teeth is broad, with low and numerous tubercles, the food of the animal is of a rather soft substance, which yields to a grinding action. Such substances are fruits, nuts, roots, or leaves, which are "triturated" and mixed with the saliva during the process of mastication. Where the vegetable food is coarse grass or tree twigs, requiring long and thorough grinding, transverse ridges of enamel are present on the cheek-teeth, as in elephants, cattle, deer, and rabbits (see Figs. 8, 17, 19). Truly carnivorous animals, which eat the raw carcases of other animals, have a different shape of teeth. Not only do they have large and dagger-like canines or "dog-teeth" as weapons of attack, but the cheek-teeth (very few in number) present a long, sharp-edged ridge running parallel to the length of the jaw, the edges of which in corresponding upper and lower teeth fit and work together like the blades of a pair of scissors. The cats (including the lions, tigers and leopards) have this arrangement in perfection (see Figs. 21 and 22). They cut the bones and muscles of their prey into great lumps with the scissor-like cheek-teeth, and swallow great pieces whole without mastication. Insect-eating mammals have cheek-teeth with three or four sharp-pointed tubercles standing up on the surface. They break the hard-shelled insects and swallow them rapidly. The fish-eating whales have an immense number of peg-like pointed teeth only. These serve as do those of the seals—merely to catch and grip the fish, which are swallowed whole.

It is quite clear that man's cheek-teeth do not enable him to cut lumps of meat and bone from raw carcases and swallow them whole, nor to grip live fish and swallow them straight off (Pl. VI). They are broad, square-surfaced teeth, with four or fewer low rounded tubercles fitted to crush soft food, as are those of monkeys (see Pl. VII and its description). And there can be no doubt that man fed originally, like monkeys, on easily crushed fruits, nuts, and roots. He could not eat like a cat.

A fundamental mistake has arisen amongst some of the advocates of vegetarianism by the use of the words "carnivorous" and "flesh-eating" in an ill-defined way. Man has never eaten lumps of raw meat and bone, and no one proposes that he should do so to-day. Man did not take to meat-eating until he had acquired the use of fire, and had learnt to cook the meat before he ate it. He thus separated the bone and intractable sinew from the flesh, which he rendered friable and divisible by thorough grilling, roasting, or baking. To eat meat thus altered, both chemically and in texture, is a very different thing from eating the raw carcases of large animals. Man's teeth are thoroughly fitted for the trituration of cooked meat, which is, indeed, as well suited to their mechanical action as are fruits, nuts, and roots. Hence we see that the objection to a meat diet based on the structure of man's teeth does not apply to the use of cooked meat as diet. The use by man of uncooked meat is not proposed or defended.

Yet, further, it is well to take notice of the fact that there are many vegetarian wild animals which do not hesitate to eat certain soft animals or animal products when they get the chance. Thus, both monkeys and primitive men will eat grubs and small soft animals, and also the eggs of birds. Whilst the cat tribe, in regard to the chemical action of their digestive juices, are so specialised for eating raw meat that it is practically impossible for them to take vegetable matter as even a small portion of their diet, and whilst, on the other hand, the grass-eating cattle, sheep, goats, antelopes, deer and giraffes are similarly disqualified from any form of meat-diet, most other land-mammals can be induced, without harm to themselves, to take a mixed diet, even in those cases where they do not naturally seek it. Pigs, on the one hand, and bears, on the other, tend naturally to a mixed diet. Many birds, under conditions adverse to the finding of their usual food, will change from vegetable to animal diet, or vice-versâ. Sea-gulls normally are fish-eaters, but some will eat biscuit and grain when fish cannot be had. Pigeons have been fed successfully on a meat diet; so, too, some parrots, and also the familiar barn-door fowl. Many of our smaller birds eat both insects and grain, according to opportunity. Hence it appears impossible to base any argument against the use of cooked meat as part of man's diet upon the structure of his teeth, or upon any far-reaching law of Nature which decrees that every animal is absolutely either fitted (internally and chemically, as well as in the matter of teeth) for a diet consisting exclusively of vegetable substances, or else is immutably assigned to one consisting exclusively of animal substances. There is no à priori assumption possible against the use as food by man of nutritious matter derived from animals' bodies properly prepared.

So far as à priori argument has any value in such a matter, it suggests that the most perfect food for any animal—that which supplies exactly the constituents needed by the animal in exactly right quantity and smallest bulk—is the flesh and blood of another animal of its own species. This is a startling theoretical justification—from the purely dietetic point of view—of cannibalism. It is, however, of no conclusive value; the only method which can give us conclusions of any real value in this and similarly complex matters is prolonged, full, well-devised, well-recorded experiment. At the same time, we may just note that the favourite food of the scorpion is the juice of the body of another scorpion, and that the same preference for cannibalism exists in spiders, many insects, fishes, and even higher animals.

Another line of argument by which some advocates of vegetarianism appeal to the popular judgment is by representing flesh-food derived from animals as something dirty, foul, and revolting, full of microbic germs, whilst vegetable products are extolled as being clean and sweet—free from odour and putrescence and from the scaremonger's microbes. This, I perhaps need hardly say, is a gigantic illusion and misrepresentation. I came across it the other day in a very unreasonable pamphlet on food by the American writer, Mr. Upton Sinclair. Putrefactive microbes attack vegetable foods and produce revolting smells and poisons in them, just as they do in foods of animal origin. It is true that on the whole more varieties of vegetable food can be kept dry and ready for use by softening with hot water than is the case with foods prepared from animals. This is only a question of not keeping food too long or in conditions tending to the access of putrefactive bacteria. It is, on the whole, more usual and necessary, in order to render it palatable, to apply heat to flesh, fish, and fowl than to fruits. And it is by heat—heat of the temperature of boiling water—applied for ten minutes or more, that poison-producing and infective bacteria are killed and rendered harmless. More people have become infected by deadly parasites and have died from cholera and similar diseases, through having taken the germs of those diseases into their stomachs with raw and over-ripe fruit or uncooked vegetables and the manured products of the kitchen garden, than have suffered from the presence of disease-germs or putrefactive bacteria in well-cooked meat. Here, in fact, "cooking" makes all the difference, just as it does in the matter we were discussing above of the fitness of flesh and bone for trituration by man's teeth.

It must be noted that the number of tubercles on the true molars may be in exceptional cases one more or one less than that given in this drawing which gives the most usual number. The word "molar" is often used to include the five cheek-teeth on each side of each jaw, but more strictly the anterior bicuspid teeth are called "pre-molars," and the three larger teeth behind them, which have no predecessors or representatives in the first or milk dentition, are called true molars or simply "molars"—a rule we have followed here.

In both upper and lower jaw we see the four incisors in the middle (Inc. 1, Inc. 2); on each side of them is the conical crown of a canine—a tooth which is greatly enlarged in the ape (see Pl. VII), but is no larger proportionately than it is here even in the most ancient known human jaw, that from the Pleistocene of Heidelberg (see "Science from an Easy Chair," Methuen, 1910, p. 405). The two small bicuspid "pre-molars" and the three large molars follow these on each side in each jaw. The crown of the most anterior (or "first") molar of the upper jaw has four cusps, tubercles, or cones on it. It is "quadri-tuberculate." The second and third molars of the upper jaw have three such prominent tubercles (excluding a row of small tubercles on the hinder margin of the second); they are, in fact, tri-tuberculate; whilst the two hindermost molars of the lower jaw have four tubercles and are called quadri-tuberculate. The first molar (M1) of the lower jaw has in this specimen five tubercles. In 60 per cent. of European lower jaws this is the case. But in 40 per cent. this tooth is quadri-tuberculate. In Polynesians, Chinese, Melanesians and negroes five tubercles are found on this tooth in 90 per cent. of the jaws examined. The apes are characterised by five tubercles on this tooth, and they are found also on the first lower molars of prehistoric men. Four tubercles only on this tooth is a departure from the ape's condition and is found more frequently in Europeans.

It is obvious that these big molar teeth, as well as the two smaller ones in front of them on each side of each jaw, are adapted for breaking up rather soft, pulpy food, and not for cutting lumps of bone or raw flesh, as are the molars of the clouded tiger (identical with those of all species of the genus Felis), shown in Figs. 21 and 22, pp. 103, 104, nor for rubbing grain, grass or herbage to a paste, as are those of the goat (Fig. 17), those of the Coypu rat (Fig. 19), and those of the elephants and mastodons.

The details of the tubercles on these molar teeth distinctly justify the conclusion that they are adapted in the two animals compared—namely, man and the gibbon—to food of the same mechanical quality, and this undoubtedly is fruit and nuts. Nevertheless such a form of tooth is equally well adapted to the texture of cooked meat, which has served many races of man for probably hundreds of thousands of years as food.

Once we remember that man is not fitted for the "raw meat" diet of the carnivora, but is fitted for the "cooked meat" diet which he has himself discovered—alone of all animals—we shall get rid of a misleading prejudice in the consideration of the question as to whether civilised men should or should not make cooked meat a portion of their diet, with the purpose of maintaining themselves in as healthy and vigorous a state as possible. Do not let us forget that ancient Palæolithic cave-men certainly made use of fire to cook their meals of animal flesh, and that probably this use of fire dates back to a still earlier period when, in consequence of this application of the red, running tongues of flame, which he had learned to produce, primitive man was able to leave the warmer climates of the earth and their abundant fruits, and to establish himself in temperate and even sub-Arctic regions.

Experiments on a large and decisive scale in regard to the value of the different foods taken by man and the question of the desirability of cooked meat as part of his diet have never been carried out, nor has the use of alcohol been studied by direct experimental method on a large scale. Inasmuch as the feeding of our Army and Navy, of prisoners, lunatics, and paupers, is the business of the State, it is obviously the duty of the Government to investigate this matter and arrive at a decision. It can be done by the Government, and only by the Government. The Army Medical Department is fully capable, and, I am told, desirous, of undertaking this investigation. Five hundred soldiers in barracks would find it no hardship, but an agreeable duty (if rewarded in a suitable way), to submit to various diets, and to comparative tests of the value of such diets. There would be no difficulty in arranging the experimental investigation. Fifty years ago similar work (but not precisely in regard to the questions now raised) was done by the Army Medical Department, under Parkes, with most valuable and widely recognised results.

FOOD AND COOKERY

Animals, taking one kind with another, nourish themselves on an immense variety of food. The flesh and the blood of other animals of all kinds, warm or cold, the leaves, twigs, fruits, juices of plants, putrid carcases, hair, feathers, skin, bran, sawdust, the vegetable mould or "humus" of the earth's surface, the sand of the sea, with its minute particles of organic detritus, all serve as food to different kinds of animals. Some are very little fettered in their tastes, and are called "omnivorous," others are bound in the strictest way to a diet consisting of the leaves of some one species of plant or the juices of one species of animal. Some of the latter class, under stress or privation, can accommodate themselves to a new food very different in character and origin from that which is habitual to them; others have no elasticity in this respect, and must have their exact habitual food-plant or food-animal, unless they are to die of starvation.

Man exhibits his great powers of accommodation to changed circumstances in respect of food as well as in other matters. If we are to suppose, as is probable, that our original ape-like ancestors fed exclusively upon fruits and an occasional egg or juicy grub, how vast are the changes in diet to which man has habituated himself! Man is sometimes said to be omnivorous, but this is not a sufficient description of the state of things which has grown up as he has spread over the earth's surface. Every race—and even many a small group of men—has its accustomed diet, to depart from which is a pain and a difficulty, even though new kinds of food may be gradually accepted and even become popular. Man has in this, as in so many other things, a large range of possible accommodation, but he has at the same time habits the continuance of which are necessary for the healthy working of the nervous system. The psychical element in the matter of food-habit is important in all higher animals, but most of all in man. The digestive organs are controlled by the nervous system, and the brain acts upon the latter in such a way as to favour or to restrain the "appetite" and the secretion of the elaborate digestive juices, so that fear, surprise, disgust, and "nausea" (that strange product of mental and physical reactions) may destroy appetite and inhibit the digestive process. There are vast populations of men who live on rice, or beans, or meal, and never eat animal food, not even milk (after babyhood), nor cheese, and would be, at a first attempt to eat it, "put off" and disgusted by a mutton chop. There are others who subsist almost entirely on fish, others who live on dried beef, others who live on the fat of whales and seals, and would be for a generation or two injured, half starved, and some of them even killed, by a change of diet. Again, there are others who consider that they must have and will be "ill" unless they had the cooked flesh of an ox or sheep as part of their daily food. Let us examine this latter group a little more fully—a group to which the nations of Europe belong, with the exception of the Italians, who are essentially a meal-, fruit-, and cheese-eating people.

Apparently at a very early time, even before the last glacial period, man had learnt the use of fire, and roasted or grilled the carcases of other animals which he killed in the chase, in order to consume them as food. We have no reason to suppose that man ever made use of the raw flesh of higher animals as his habitual diet. His teeth are not, and never were, from his earliest ape-like days, adapted to true carnivorous diet. Cooked meat is not the food of a carnivor, but is an adaptation of the flesh of animals to the requirements of a frugivorous animal. Probably the use of grain and cultivated vegetable food is a later step in human progress than the roasting of meat. The Neandermen, and even the later Reindeer-men (Cromagnards), had no cultivated fields, but lived on roasted meat (of beasts, birds, and fish) and wild fruits. We know how thoroughly the most ancient Greeks enjoyed the long slices of roasted meat cut from the chine, as told in the Homeric poems, and everywhere in Europe after the neolithic or polished-stone period, meat was a main article of diet, in conjunction with the vegetable products of agriculture. In this country, after the Norman conquest, meat-eating was greatly favoured by the important industry which grew up in hides. The land was well suited for the pasturage of cattle, and owing to the smallness of the population and the abundance of cattle slaughtered for their hides, meat was almost to be had for the asking. It was thus that Englishmen became great meat-eaters and that "the roast beef of Old England" was established. Later the same superfluity of meat—in this case, "mutton"—recurred and became general when wool-growing and the manufacture of woollen goods developed into important industries. Relatively to the population there was more "meat" of oxen and sheep in this country than on the continent of Europe, and this disproportion has been maintained.

But the increase of population has led to a considerable change in the diet of a very large proportion—the poorer part—of the community. Whilst the families of the better-paid working class and all the middle and upper class continue to eat meat, the agricultural labourer and the poorer workmen in towns live chiefly on flour, sugar, bacon, and cheese. Probably they have become habituated to this diet, and, provided that the quantity is sufficient, it cannot be maintained that the diet, in which meat is nearly or altogether absent, is unhealthy. Many vigorous and muscularly well-developed populations in other lands thrive on exclusively vegetable food.

A curious and not altogether comforting reflection is that if the inexpensive and simple food of the agricultural labourer is sufficient, the section of the community which spends from five to ten shillings per head a day on a mixed diet of meat, fish, eggs, and vegetables is guilty of waste and excess. Here, however, the remarkable, and, in fact, exceptional domination of "habit" (in the case of man), in regard to both the actual articles of food and the mode of its preparation, has to be recognised. Such and such inexpensive and unskilfully prepared food may contain more than the necessary amount of proteids (that is, matters like flesh, the casein of cheese and of vegetables, and the albumen of eggs), of hydro-carbons (i.e., fats), of carbo-hydrates (i.e., starch and sugar), yet if you were suddenly to compel a man accustomed to well-cooked meat to live on such food he would be unable to assimilate it, his digestive organs would refuse to work, and he would become, if not seriously ill, yet so ill-nourished and sickly that he would be unfit for his work and readily fall a victim to disease. It is, in fact, impossible to lay down any scheme of diet based on the mere provision of the necessary quantities of food materials whilst ignoring the formed habits of the individual and the relation of the psychical conditions which we call "taste," "appetite," "fancy," "disgust," to the actual processes of digestion and the consequent efficiency of the proposed diet.

No doubt gradually, after a few generations, a whole people may become healthily habituated to a diet which would have been positively injurious to their forebears, and no doubt individuals may be led by fortitude or by necessity in time (perhaps weeks, perhaps years) to acquire a tolerance, or even enjoyment, of food at first repulsive, and therefore injurious. The difficulty in the matter is not that of correctly determining what is physiologically sufficient for the human animal, nor even what would be a healthy diet for a community when once, after a transition period of distress and injury, habituated or "attuned" to that diet. The difficulty is to arrive at a conclusion as to what is really the suitable and reasonable diet for an individual—yourself or one like yourself—having regard to the lifelong habits of the individual, and the consequent nervous reactions established in him or her in relation to the taste, quality, and mode of presentation of food. Robust people, so long as they get what suits their own uncultivated taste, are apt to make very light of what they call "fancies" about food, and to overlook their real importance.

Feeding on the part of civilised man is not the simple procedure which it is with animals, although many animals are particular as to their food and what is called "dainty." The necessity for civilised man of cheerful company at his meal, and for the absence of mental anxiety, is universally recognised, as well as the importance of an inviting appeal to the appetite through the sense of smell and of sight, whilst the injurious effect of the reverse conditions, which may lead to nausea, and even vomiting, is admitted. Even the ceremonial features of the dinner table, the change of clothes before sitting down to the repast, the leisurely yet precise succession of approved and expected dishes, accompanied by pleasant talk and light-hearted companionship, are shown by strict scientific examination to be important aids to the healthy digestion of food, which need not be large in quantity, although it should be wisely presented.

These psychical conditions of healthy feeding are not trivial matters, as we are too apt to suppose. They are part, and a very important part, of the physiology of nutrition, and so deserving of scientific inquiry and of practical attention. They have been made the subject of careful experiment by a Russian physiologist, Pavloff. At a recent meeting of the British Association this matter was brought under discussion in the Physiological Section, and it was pointed out by the author of a very interesting communication that the whole question as to what is and what is not a sound and healthy diet is too often dealt with by writers who ignore the psychical (or shall we say the cerebral?) factor. Cases were cited of dangerous arrest of the power of digesting, or even of swallowing, food which were cured by giving the patient some apparently inappropriate and probably harmful article of food for which he or she had a fancy, such as a grilled salmon-steak, the last thing which would be spontaneously recommended by a medical man to a patient who had been suffering for weeks from inability to take food. The willingness is all—the assent, the approval of the cerebral centres, and the consequent unlocking of the whole arrested mechanism of digestive secretions and movements. Such a case is only an extreme instance. But it is undoubtedly the fact that just as the sight of so small a thing as a drop of blood, or even the word "blood," will on occasion cause a strong, healthy man to faint, so quite a small excess or defect in the accustomed quality of food will at times arrest the appetite and digestive processes of a healthy man. To many a healthy individual one among many flavours and savours associated with agreeable food is necessary in order that healthy appetite and proper digestion may be set going, and the absence of the right flavour and the presence of what is, in his experience, a wrong and disgusting smell or taste in the food set before him, will produce nausea and complete arrest of the digestive processes.

It is apparently owing to this cause that "tinned meats" have proved to be of little value as rations for an army in campaign, for exploring expeditions, and for remote mining camps. It is not that such tinned meats do not contain the necessary constituents of food, or that they contain poisonous substances, but that they produce a sense of disgust, and arrest the digestive processes. Soldiers, travellers, and miners have assured me that they prefer a dry biscuit and dried, or salted, or sugared meat, to the supposed more "tasty" tinned meats, and that such is the general experience of their comrades.

Of similar nature is another very serious trouble, in regard to the healthy feeding of the modern Englishman, which has come upon us in consequence of the quite modern system of huge restaurants, whether in London or in the very large hotels, which are now run in Swiss, Italian and English summer resorts. Hundreds of visitors are "catered for" daily. There is no attempt at anything which deserves the name of cookery. Great monopolists control the supplies, and contract to deliver to these hotels, even in out-of-the-way localities, so much ice-stored, "mousey" fish, "mousey" quails, stringy meat, impossible vegetables and fruits, gathered from the cheapest markets of Europe and of a quality just not bad enough to cause a revolt among the hotel visitors. The heating of the food is done by patent machinery in ovens and by the use of boiling fat. No cook is in these circumstances possible, with his artistic feeling for the production of a perfect result of skill and taste. A kind of bottled meat-flavoured sauce, manufactured from spent yeast, is used to make the soups, and is poured, with an equally nauseating result, over the hard veal, the tough chicken, the "mousey" quails, and the tasteless beef and mutton, which are never roasted, but are baked or stewed in boiling fat—though shamelessly described as "rôtis" in the pretentious and mendacious "menu" placed on the dinner-table. The consequence is that the tourist, who has been overfed at home, eats very little, and his health benefits. But in such an hotel the man who lives carefully when at home, and desires a simple but properly cooked meal, is reduced to a state of indigestion, semi-starvation and misery.

The Englishman who is disgusted by the new mechanical methods of cookery in the great hotels of Continental "resorts," returns to London, and finds the same atrocious system at work—not only in the public restaurants, but in his club. Nowhere in London can you rely on being served with really fresh fish, however highly you may pay for it. Rarely it is fresh, usually it is not. The ice storage people take good care that you shall not obtain fresh fish, and so retain your taste for it. Nowhere at club or restaurant, with rare exceptions, can you obtain meat roasted in the old-fashioned way on a roasting-jack, carefully "basted" during the process, and served when exactly cooked to a turn. There were, only a few years ago, one or two such places surviving—both clubs and restaurants—where proper roasting was done, but, like the rest, they have now adopted lazy, economical, money-saving methods. Their managers calculate that what they do will serve. It is good enough for the crowd! So at last you abandon the efforts to obtain decent simple food, in club or hotel, and dine with your friend en famille. The same thing confronts you. The joint has been baked in an oven, of which it smells, and is surrounded by a sickly gravy, produced by pouring hot water over it! In conversation with your hostess, you find that she knows nothing whatever about the simplest elements of the preparation of food. She tells you she avoids roasting because it necessitates a large fire and an extra expenditure of £5 a year on coal, and she also purchases those mouldy, frost-bitten potatoes instead of the best, because they cost half as much as sound ones—and she herself does not care for potatoes. They are fattening!

Sometimes at a restaurant or club, served by a foreign "chef," a Yorkshire pudding, as hard as a stale loaf of bread, is handed round in slabs with the so-called "roast" beef. It is not roasted: it is baked beef, and the pudding is an ill-tasting baked mess, also. Nowhere in London in public or private house do I ever see the properly cooked article. True Yorkshire pudding can only be made by placing it under the roasting joint, which drips digestion-promoting essences into the pudding whilst itself rotating, hissing and spluttering—as did the joints roasted in the caves long ago by the prehistoric Reindeer-men. The scientific importance of good roasting and grilling is that a savour is thereby produced which sets the whole gastric and digestive economy of the man who sniffs it and tastes it, at work. Possibly our successors, a generation or two hence, will have learnt to do without this, and will have acquired as intimate and happy a gastronomic relation to what now are for us the nauseous flavours of superheated fat (rarely renewed), and of the all-pervading gravy fabricated by chemical treatment of yeast, as that which we ourselves have acquired in regard to the old-established and painstaking cookery of the early Victorian and many preceding ages.

Medical men who are occupied as specialists with the study of very young children have clearly demonstrated that the implanting of tastes, tendencies and habits in infants of from two to eight years of age has an immense importance in their subsequent development. Character and capacity are really formed in those early years. Food preferences, no less than mental and moral qualities, are then created. Yet the children of both rich and poor are in these early stages either left to haphazard or entrusted to ignorant nursemaids. For those of us who were not born to the present system the transition to the new methods of wholesale cookery is an abomination, and to escape from them a matter of difficulty. We have to secure an ancient roasting-jack and a large clear fire in our own kitchen, and to instruct our cook—since no woman has taught her what she ought to know—in the art of roasting and grilling, in the preparation of Yorkshire pudding, in the mystery of the marrow-bone and the proper and distinct use of garlic, onions, shalots, chives, chervil, tarragon, marjoram, basil, other herbs, and divers peppers, and finally to train her in the supreme accomplishment of the seasoning of a salad.

Maybe that the present established relations of our appetites to the time-honoured savours, by which the ancient Jews sought to propitiate the Deity, are destined to be superseded. On the other hand it is quite possible that all the juggling of modern "machine" cookery is a false step, and injurious to digestion and health. It is not unlikely that there is no relish which has so sure a hold on the digestion of European man, no appeal to the cerebral mechanism controlling the liberation of his gastric juices, which is so infallible as that emanating from "well and truly" roasted or grilled meat.

It is not easy to account for the present neglect of decent cookery and the triumph of the sham French cookery (for it is not French at all!) which is at present foisted on a long-suffering public. Probably the enormously increased number of visitors to foreign resorts and of frequenters of restaurants in London have led to huge enterprise in "catering," and to a monopoly which has driven out of existence the smaller establishments, where alone the artist-cook can flourish. But it seems that the neglect of decent cooking is also due in this country to a racial incapacity and indifference which leads both men and women to despise "taking pains" about small things, and brings them into the world devoid of the desire to carry out with skill those small enterprises on which much of the sweetness and gaiety of life depends.

Even in the time of Charles II the skill and seriousness of French cookery as compared with our own was recognised. The high reputation of Scotch cooks at the present day seems to be due to an inheritance of traditions from the days of close association of the Scotch and French Courts. Up to nearly 100 years ago roasting was as usual a method of cooking meat in Paris as in London. There were "rôtisseries" in Paris in the old days. High prices and thrift have led to the decadence of roasting as a popular method of cooking meat in France, but the great "chef" in a private house in Paris still produces the most perfect roast beef and roast saddle of mutton (better than you will find in England) in the old-fashioned way. So indifferent, or perhaps hopeless, are Englishmen in regard to cookery that they drink a strong champagne throughout dinner, content to drown the insipid taste of the food in the fine flavour of a drink upon which they can rely. An Englishman dining at a first-rate restaurant will usually spend twice as much for wine as for food, whilst a Frenchman will reverse the proportions. Another difference is one for which women are responsible. In Paris a party of French men and women at a table in a good restaurant enjoy their food, laugh and talk with one another, and do not concern themselves with the company at other tables. It would be bad manners to do so. But English-speaking women, when dining in public, seem to be chiefly interested, not in their food nor in their own party, but in pointing out to one another the celebrities or notorieties or eccentricities seated at other tables. So long as the place is fashionable and noisy, the food is negligible and neglected.

For some reason, which I am unable to discover, the women of England (it is not the case with those of France and Germany) have, with rare exceptions, no interest in or liking for "cookery," and yet the men have left the management of it entirely in their hands. Male "chefs" of English nationality are rare specimens, though they are, as a rule, the best at grilling and roasting. On the other hand, in France, where women no less than men value and understand cookery, there is an enormous body of professional male cooks. English-women of means and education have to such a degree neglected all knowledge of cookery and of the quality and criticism of kitchen supplies, such as meat, fish, birds, and vegetables, that there is no one to teach the poor country girls (who become cooks in the majority of households) the elements of the very difficult and important duties which they are expected—in virtue of some kind of inspiration or native genius—to discharge with skill and judgment: nor is there any head of a household capable of seeing that the necessary care and trouble are given. It is wonderful, under the circumstances, how clever and willing our domestic cooks are. A considerable section of English middle-class women at the present day are allowed by the men, who should guide them so as to make them honourable and useful members of the community, to grow up in complete ignorance of the essential parts of the art of cookery. This was not the case a hundred years ago. Now a large proportion of them have been led by bad example and foolish notions to give up such matters to "the servants," whether they are able to afford competent servants or even to judge of the competence of a servant or not. Many of these "mistresses" now devote themselves exclusively to "dress," "amusements," "charity," "politics," and dabbling inconsequently in various crazes. They are not to blame. It is the men who are to blame who deliberately neglect to give to their womenkind a training and education which shall make them real mistresses of household arts and business, so that they may be thus filled with the happy conviction (which is the one thing they most desire and most often cannot gain) that they are of real use—are really wanted—in the world.

In conclusion, let me tell of a great German sports-man, Major von Wissman, Governor of German East Africa, now no more, who came to see me at the Museum nine years ago. It was his first visit to London, and I took him to lunch at a famous grill-room. Happily, though roasting is dying out, the art of grilling still survives in this country, but nowhere else in

Europe. Von Wissman said—"Can I have beer where we are going?" "Yes, certainly," I said. "German beer?" he asked. "No," I replied. "Something much better." When we were seated, I ordered a pint tankard of Reid's London stout for my friend. It was in perfect condition. He put his lips to it in doubt, but did not remove them until, with reverential drooping of the eyelids, he had emptied the tankard. "The very finest beer I have ever swallowed," he said. "What in the name of goodness is it?" I told him, and ordered him more. Soon a perfectly grilled chop and a large, clean, floury potato were before him. He proceeded to eat, and was really and unaffectedly astonished. "But this is marvellous," he said, "wonderful! enchanting! I have never really tasted meat before in my life. Reitzend! Colossal!" He had a steak to follow, and I was pleased to have been able to show him something which I knew (by experience of that city) they could not produce in Berlin. Three days later I went over to the same hospitable grill-room for a chop, and told the gifted grill-cook (the French, in former centuries, had a proverb, "Anyone may learn to be a cook, but one must be born a 'rotisseur'") of the admiration he had excited in the Emperor William's friend. "Yes, sir," he said, "I fancy he did like it, for he came here by himself yesterday and the day before, and took the same grills and stout." Von Wissman was staying at the German Embassy, but was drawn all the way to South Kensington by the sweet savour of the grill-room—an instance of what the physiologists call "positive chemotaxis."

What I have here written on food and cookery is no "gourmet's" praise of indulgence in the pleasures of the table, nor is it an expression of a mere personal preference. It is a protest, based on scientific grounds, against the neglect of one of the bulwarks of health—the honest traditional cookery which flourished in London forty years ago.

SMELLS AND PERFUMES

The old saying, "De gustibus non disputandum," is based upon the fact that both liking and the repulsion evinced by human beings for different odours (including those odours which we call flavours) are not matters of general agreement. Thus the smells of garlic and of onions, and even of assafœtida, are to many men among the most attractive and appetising in existence—to very many they are, on the other hand, repulsive. High game, a certain kind of putrid fish ("Bombay ducks"), and again rotten cheese are attractive to many men and offensive to as many more. Many animals revel in the smell and flavour of carrion, and even of manure, which they devour. There are well-known flowers which attract insects, not by the possession of the sweet perfumes appreciated and extracted by mankind, but by a smell like that of putrid meat, which so far misleads blue-bottle flies as to cause them to lay their eggs on the reeking blossom. So diverse are the tastes of men and animals in these matters that it is remarkable when we find agreement among them, as, for instance, in the attraction for butterflies of those delicate scents which also are agreeable to ourselves in such flowers as the rose, the jasmine, the heliotrope and the honeysuckle.

There seems to be no rule or principle at work by which smells can be definitely classed as either pleasant or unpleasant. Even perfumes carried by some of the inhabitants of Western Europe with the intention of making themselves attractive to their fellow-citizens are often repulsive to a certain proportion of those who come near them, as, for instance, is the case with the extract of the East Indian herb "patchouli." In regard to our other senses there is a general agreement amongst mankind, which extends also to all animals, as to what is agreeable and what is disagreeable. There are definite mathematical laws as to harmony and melody in sound and colour which affect animals and ourselves to a large extent similarly. Sweets are agreeable and bitters are disagreeable, though it is the fact that the snail, which loves sugar, recoils from saccharine, and there are "mites" (Acari) which feed with avidity on bitter strychnine! Excess of heat and of cold is disliked by animals and all men, whilst the sense of touch is pleasurably or painfully affected in much the same way in most men and animals, more than is the case with regard to any other of the senses. The sense of smell depends upon immediate and personal experience of "association" for the determination of pleasure or pain, attraction or repulsion, as the result of its being called into operation. It is a very general experience that odours are more efficient in arousing memory than are mere colour effects or sounds. Not only in animals with acutely developed olfactory powers, but also in man, an odour—a peculiar perfume—will start a whole chain of reminiscence when sight and sound have failed to do so. It is due to this close association with memory (conscious or unconscious) that an odour is agreeable or disagreeable.

In itself an odour is neither attractive nor repulsive. The acrid fumes of sulphur, chlorine, ammonia, and such bodies are not simply "odours" but corrosive chemical vapours, which act painfully upon the nerves of common sensation within the air-passages of the nose and throat and not exclusively, if at all, on the terminations of the olfactory nerves. An odour—that which acts on the special nerves of smell distributed in chambers of the nose—acquires its attractive or its repulsive quality only as the result of mental association with what is beneficial (suitable food, mates, friends, safety, home, the nest), or with what is injurious (unsuitable food, poison, enemies, danger, strange surroundings, solitude). Hence it is intelligible that the man accustomed to garlic or onions in his food is strongly attracted by their smell. So too the man whose tribe or companions have learnt by necessity to eat slightly putrid meat, fish, and cheese is attracted by their odour, though for others these odours are associated rather with what is poisonous and injurious. The dislike of the smell of sewer-gas and foul accumulations of refuse was not known to former generations of men (even in European cities a couple of hundred years ago) any more than it is to-day to the more unfortunate poorer classes, to many modern savages, to hyenas, and several other animals and birds which inhabit lairs and caves which they make foul. The odour of putrescence has become actually painful and almost intolerable to the more cleanly classes of mankind, owing to the association with it, as the result of education, of fear of disease and poisoning. Either conscious or unconscious association of an odour with what is held, either as the result of tradition or through personal experience, to be beneficial and of pleasant memory, or, on the contrary, injurious and of painful connection, determines man's liking for and choice or rejection of, odours and flavours. One can account with fair success on this basis for one's own preferences and dislikes in the matter.

On the other hand, odours exist in vast variety amongst plants and animals which have not acquired any special association or significance. We find that some organisms produce as a result of their chemical life material which oxidises and gives out light and so these organisms are "phosphorescent" without any consequence, good or bad, to themselves. And then we come upon others (as, for instance, the glow-worms and fire-flies) which have made use of this "accidental" quality, and produce phosphorescent light in special organs so as to attract the opposite sex. Again, we find that the red-coloured oxygen-seizing crystalline substance hæmoglobin exists in the blood of a vast number of animals, and might as well be green or colourless for all the good its colour does them. Yet here and there the splendid red colour which this chemical gives to the blood becomes of great importance as a "decoration," or "sex-ornament." The comb of the domestic fowl, the wattles of the turkey, but above all the supreme beauty of the human race—the cherry-red lips and the crimson-blushing cheek of healthy youth—owe their wonderful colour to the red blood which flows through them. So at last the redness, of the oxygen-carrier is turned to account. So it must be also with odorous substances. Many have been called into existence, but few have been chosen in the long course of animal evolution and selected as the important means of repulsion or attraction.

There are odorous substances attached to many of the lower animals which seem to have no significance, but just happen to be the result of necessary chemical changes, not aimed (so to speak) at their production. Of course, it is very difficult to form a certain and definite conclusion as to their uselessness as odours. For instance, nearly all the sponges when fresh and filled with living protoplasm have a curious smell which reminds one of that given off by a stick of phosphorus. Marine sponges have it and so has the beautiful green or flesh-coloured liver sponge (common on the wood of rafts and weirs in the Thames). A rather uncommon marine worm, called Balanoglossus or the acorn worm, has a very strong and unpleasant smell like that of iodoform. In neither case is the nature of the odorous body known, nor its use to the animal suggested. Smelts smell like cucumbers: the green-bone fish and the mackerel smell alike. One of the common earth-worms has a strong aromatic smell, and the common snail, as well as the sea-hare and one of the cuttle-fishes (Eledone), smells like musk. Musk itself is produced, as a scent attracting the opposite sex, by several animals—musk-deer, musk-sheep, musk-rats. I am not now attempting to enumerate the well-recognised odours of animals such as are extracted from them by man in order to "opsonize" himself, but am pointing to the more obscure cases. There is not a very great or marked variety in the odours of fishes; but reptiles with their dry, oily skins give off various aromatic smells, none of which are valued by man. Toads have distinct odours, and one kind

(Pelobates fuscus, or the heel-clawed toad), common in Europe, but not British, is known locally as the garlic toad on account of its smell. There are amongst carnivorous mammals various smells allied to that of civet which are not so agreeable to man as that substance; for instance, the odour of the fox and of the badger, and yet more celebrated, the terrible, awe-inspiring smell of the fluid emitted in self-defence by the skunk from a sac in the hinder part of the body. Horses, cows, goats, sheep, and the giraffe have their distinctive odours. Many of the herbivorous animals secrete a colourless fluid from large glands opening on or near the feet, and also from a gland in front of the eye (similar glands occur in other strange positions), which has not a smell familiar to man—that is to say, not one which has been recognised and described—yet seems to be readily "smelt" by the animals of its own kind. The bats—especially the large frugivorous bats—have a very unpleasant, frowsy smell.

An important fact about animal smells is that many which we might be inclined to attribute to the animal which diffuses them, are really due to the fermentative or putrefactive action of bacteria which swarm on the skin and in the intestines of animals. It is often difficult to decide how far a peculiar animal odour is due directly to a substance secreted by the animal, and how far the odour of that substance is modified or even entirely produced by the chemical changes set up in secretions of the body-surface by bacteria. Several distinct repulsive smells liable to occur on the human body are due to want of cleanliness in destroying bacteria by proper antiseptics. The fatty and waxy secretions of the skin are often decomposed by bacteria, even before complete extrusion from the glands in which they are formed, whilst the decomposition of food in the mouth and intestines by bacteria alters materially both the natural odour of the animal's breath and the smell of the intestinal contents. In young and healthy animals in natural conditions there is some check—it is not easy to say what—upon the putrefactive activities of the omnipresent bacteria. The skin of a healthy young animal has a pleasant odour, and its breath (notably in the case of the cow and the giraffe) is naturally sweet-smelling. The same should be the case, under perfectly healthy conditions, with human beings.

There is one important cause of animal odours and flavours upon which I have not hitherto touched. Many animals acquire an odour or flavour directly from the food upon which they feed. Certain odorous bodies are in the food and are taken up into the blood of the consuming animal unchanged, and are then thrown out by secreting glands on the skin. This is the case with the odorous substance of onions. People do not smell of onions after they have eaten them in consequence of particles of onion remaining in the mouth. The volatile odoriferous matter of the onion is absorbed into the blood. It passes out first through the lungs and later through the small fat-forming glands in the skin. It is difficult to ascertain how far animals derive their odours in this way in a complete state from their food, and how far they chemically construct them afresh by their own activity. No doubt both processes occur; but in plants the odorous bodies are built up entirely by the chemical action of the plant itself upon simple salts of carbonic acid, ammonia and nitrates. Animals can certainly take highly elaborated chemical bodies into their digestive organs without destroying them and absorb them unchanged into the blood and deposit them in the tissues. Thus the canary is made to take up the red colour of cayenne pepper and deposit it in the feathers. Thus the green oysters of Marennes acquire their colour from minute blue plants (diatoms) on which they feed. And thus, too, the canvas-backed ducks of the United States take into their tissues the odorous matter of celery, and our own grouse the flavour of heather, whilst fish-eating birds and whales in this way acquire a fishy taste. So, too, the flounders and the eels of the Thames, and even salmon in muddy rivers, acquire a taste like the smell of river mud. It is probable that many of the odours of animals (but by no means all) are thus derived directly from their food, or are produced by very slight changes of the odorous bodies absorbed in food. Mutton and beef owe their savour in some degree to the scents of the grasses on which sheep and oxen feed. And it is not improbable that the sheep-like smell which the Chinese detect in the European, comes to the latter direct from his general use of the sheep as food.

Plants are the great chemical manufacturers in the world of life, and second to them come our human industrial and scientific chemists. And though we must claim for animals some power of manufacturing distinct odorous bodies from inodorous nutritive matter assimilated by them, it is probable that in many cases the odour which is characteristic of an animal is derived by no very complicated change from odorous bodies existing in its habitual food.

A curious case of a substance valued as perfume by civilised man, and yet coming from a source whence sweet odours would hardly be expected, is that which is known as "ambergris," or "ambre gris" (grey amber). It is still used in the manufacture of esteemed perfumes, and is sold at five guineas the ounce. It is found floating on the surface of the ocean, and is a concretion of imperfectly digested matter from the intestine of a whale— probably the sperm-whale. It is a grey, powdery substance, and in it are embedded innumerable fragments of the horny beaks and sucker-rings of cuttle-fishes—creatures which form the chief food of the sperm-whale and other toothed whales. I have already mentioned above that one of our common cuttle-fishes (the Eledone moschata) has a strong odour of musk, and it is possible that ambergris owes its perfume to the musk-like scent of the cuttle-fish eaten by the whale in whose intestine it is formed. Another "smell" which is extremely mysterious is that produced by two quartz-pebbles, or even two rock-crystals, or two pebbles of flint or of corundum, when rubbed one against the other. A flash of light is seen, and this is accompanied by a very distinct smell, like that given out by burning cotton-wool. It is demonstrated—by careful chemical cleaning before the experiment—that this is not due to the presence of any organic matter on or in the stones or crystals used. It seems to be an exception to the rule that "odours" (as distinct from pungent vapours or gases) are only produced by substances formed by plants or animals. Perhaps that is not so completely a rule as I was inclined to think. It is true that one can distinguish the "smells" of chlorine, of bromine, and of iodine from one another. And there are statements current as to the distinctive smells of metals—though they may possibly be due to the action of the metals on organic matter. In any case it seems, according to our present knowledge, that the smell given out by the rubbing of pieces of silica (quartz, flint, etc.) is due to particles of silica (oxide of silicon) volatilised by the heat of friction, which are capable of acting specifically on the olfactory sense-organ.

KISSES

"Among thy fancies, tell me this,
What is the thing we call a kiss?
I shall resolve ye what it is."
—Robert Herrick

Kissing is an extremely ancient habit of mankind coming to us from far beyond the range of history, and undoubtedly practised by the remote animal-like ancestors of the human race. Poets have exalted it, and in these hygienic days doctors have condemned it. In the United States they have even proposed to forbid it by law, on the ground that disease germs may be (and undeniably are in some cases) conveyed by it from one individual to another. But it is too deep-rooted in human nature, and has a significance and origin too closely associated with human well-being in the past, and even in the present, to permit of its being altogether "tabooed" by medical authority.

There are two kinds of "kissing" practised by mankind at the present time—one takes the form of "nose-rubbing"—each kiss-giver rubbing his nose against that of the other. The second kind, which is that familiar to us in Europe, consists in pressing the lips against the lips, skin, or hair of another individual, and making a short, quick inspiration, resulting in a more or less audible sound. Both kinds are really of the nature of "sniffing," the active effort to smell or explore by the olfactory sense. The "nose-kiss" exists in races so far apart from one another as the Maoris of New Zealand and the Esquimaux of the Arctic regions. It is the habit of the Chinese, of the Malays, and other Asiatic races. The only Europeans who practise it are the Laplanders. The lip-kiss is distinguished by some authorities as "the salute by taste" from nose-rubbing, which is "the salute by smell." The word "kiss" is connected by Skeat with the Latin "gustus," taste; both words signify essentially "choice." But it would be a mistake to regard the lip-kiss as merely an effort to taste in the strict sense, since the act of inspiration accompanying it brings the olfactory passages of the nose into play. Lip-kissing is frequently mentioned in the most ancient Hebrew books of the Bible, and it was also the method of affectionate salutation among the Ancient Greeks. Primarily both kinds of kissing were, there can be no doubt, an act of exploration, discrimination, and recognition dependent on the sense of smell. The more primitive character of the kiss is retained by the lovers' kiss, the mother's kissing and sniffing of her babe, and by the kiss of salutation to a friend returning from or setting out on a distant journey. Identification and memorising by the sense of smell is the remote origin and explanation of those kisses. The kissing of one another by grown-up men as a salutation was abandoned in this country as late as the eighteenth century. "'Tis not the fashion here," says a London gentleman to his country-bred friend in Congreve's "Way of the World." But we have, most of us, witnessed it abroad, and perhaps been unexpectedly subjected to the process, as I once was by an affectionate scientific colleague. Independently of the more ordinary practice of kissing—there is the "ceremonial kiss"—the kissing of hands, or of feet and toes, which still survives in Court functions—whilst the Viennese and the Spaniards, though they no longer actually carry out their threat, habitually startle a foreigner by exclaiming—"I kiss your hands." The Russian Sclavs are the most profuse and indiscriminate of European peoples in their kissing. I have seen a Russian gentleman about to depart on a journey "devoured" by the kisses of his relations and household retainers, male and female. Among the poor in rural districts in Russia this excessive habit of kissing leads to the propagation of the most terrible ulcerative disease among innocent people—as related by Metchnikoff in the lectures on modern hygiene which he gave in London some seven or eight years ago (published by Heinemann).

We may take it, then, that the act of kissing is primarily and in its remote origin an exploration by the sense of smell, which has either lost its original significance, and become ceremonial, or has, even though still appealing to the sense of smell, ceased to be (if, indeed, it ever was so) consciously and deliberately an exercise of that sense. This leads us to the very interesting subject of the sense of smell in man and in other animals. There is no doubt that the sense of smell is not so acute in man as it is in many of the higher animals, and even in some of the lower forms, such as insects. It is the fact that so far as we can trace its existence and function in animals, the sense of smell is of prime importance as distinguishing odours which are associated either with objects or conditions favourable to the individual and its race, or, on the other hand, hostile and injurious to it. It never reaches such an extended development as a source of information or general relation of the individual to its surroundings as do the senses of sight, hearing and touch. It depends for its utility on the existence of odorous bodies which are not very widely present, and are far from universal accompaniments of natural objects. Apart from some pungent mineral gases, all odorous bodies are of organic origin. Even as recognised by the less acute olfactory sense of man, the number and variety of agreeable and of disagreeable scents, produced by various species of animals and plants, is very considerable. But there is no doubt that the number and variety discriminated by such animals as dogs and many of the other hairy, warm-blooded beasts is far greater. The nature of the particles given off by odorous bodies which act on the nerve-endings of the organs of smell of animals, is remarkable. They are volatile; that is to say, they are thrown off from their source and float in the air in a state of extreme subdivision. Unlike the particles which act upon the nerves of taste, they are not necessarily soluble in water, and though often spread through and carried by liquids, are in fact rarely dissolved in water. The dissolved particles which act upon the nerves of taste can be distinguished by man into four groups—sweet, sour, bitter, and saline. But no such classification of "smells" is possible. As a rule mankind confuses the "taste" of things with their accompanying "smell." The finer flavours of food and drink not included in the four classes of tastes are really due to odoriferous particles present in the food or drink, which act on the terminations of the olfactory nerves in the recesses of the nose, and excite no sensation through the nerves of taste.

The part of the brain from which the nerves of smell arise is of relatively enormous size in the lower vertebrates—as much as one fifth of the volume of the entire brain in fishes—a fact which seems to indicate great importance for the sense of smell in those forms. Even in the mammals (the hairy, warm-blooded, young-suckling beasts) the size of the olfactory lobes of the brain and of the olfactory nerves, and the labyrinthine chambers of the nose on which the nerves are spread, is very large, as one may see by looking at a mammal's skull divided into right and left halves. And it seems immoderately large to us—to man—because, after all, so far as our conscious lives are concerned, the sense of smell has very small importance. Yet man has a very considerable set of olfactive chambers within the nostrils and has large olfactory nerves. Not rarely men and women are found who are absolutely devoid of the sense of smell, and the same thing occurs with domesticated cats and dogs. In these cases the olfactory lobes of the brain are imperfectly developed. It is found that men in this condition suffer but little inconvenience in consequence. We are able, through their statements, to ascertain what parts of the savoury qualities of food and drink belong to taste and what to smell. Such individuals do not perceive perfumes, the bouquet of wine, or the fragrance of tobacco, nor can they appreciate the artistic efforts of a good cook. But they are spared the pain of foul smells, and possibly in this way they may incur some danger in civilised life through not being able to detect the escape of sewer-gas or of coal-gas into a house, or the putrid condition of ice-stored fish, birds, and meat. A friend of my own, who is devoid of the sense of smell, inherited this defect from his father, and has transmitted it to some of his children. I was surprised to find in conversing with him how often I alluded to smells, either pleasant or unpleasant, when (as we had agreed he should) he would interrupt me and say that my remark had no meaning for him.

Some have a far more acute sense of smell than others, and again some men, probably without being more acutely endowed in that way, pay more attention to smells, and use the memory of them in description and conversation. Guy de Maupassant is remarkable as a writer for his abundant introduction of references to agreeable and mysterious perfumes, and also to repulsive odours. But some men certainly have an exceptionally acute sense of smell, and can, on entering an empty room, recognise that such and such a person has been there by the faint traces—not of perfumery carried by the visitor—but of his individual smell or odour. This brings us to one of the most important facts about odorous bodies and the sense of smell, namely, that not only do the various species of animals (and plants) each have their own odour—often difficult or impossible for man, with his aborted olfactory powers, to distinguish—but that every individual has its own special odour. As to how far this can be considered a universal disposition is doubtful. It is probable that the power of discriminating such individual odours is limited (even in the case of dogs, where it is sometimes very highly developed), to a power of discriminating the distinctive smells of the individuals of certain species of animals, and not of every individual of every species. Everyone knows of the wonderful power of the bloodhound in tracking an individual man by his smell, but dogs of other breeds also often possess what seems to us extraordinary powers of the kind. On a pebbly beach I pick up one smooth flint pebble as big as a walnut. It is closely similar to thousands of others lying there. I hold it in my hand without letting my fox-terrier see it, and then I throw it. It drops some eighty yards off among the other pebbles, and I could not myself find it again. But the dog runs forward, notes vaguely by ear and by eye the spot where it strikes, and then commences a systematic circling within about ten yards of the spot. In half a minute he pounces with the utmost assurance on to one selected stone, and brings it to me. It is invariably the stone which had been in my hand, unseen by the dog, thrown by me, and detected by the smell I have communicated to it.

Not only is the discrimination of individuals by the sense of smell a very astonishing thing, but so also is the obvious fact that the total amount of odoriferous matter which is sufficient to give a definite and discriminative sensation through the organ of smell is of a minuteness beyond all calculation or conception. These two facts—the almost infinite individual diversity of smell and the almost infinite minuteness of the particles exciting it—render it very difficult to form a satisfactory conclusion as to the nature of those particles. It has been from time to time suggested that the end organs of the olfactory nerves may be excited, not by chemically active particles, but by "rays," olfactive undulations comparable to those of light. Physicists have not yet been able to deal with the problem, but the recent discoveries and theories as to radio-active bodies such as radium may possibly lead to some more plausible theory as to the diffusion and minuteness of odorous particles than any which has yet been formulated. An example of the minuteness of odoriferous particles is afforded by a piece of musk which for ten years in succession has given off into the changing air of an ordinary room "particles" causing a readily recognised smell of musk, and yet is found at the end of that time to have lost no weight, that is to say, no weight which can be appreciated by the finest chemical balance. An analogy (I say only an analogy, a resemblance) to this is furnished by a pinch of the salt known as radium chloride, no bigger than a rape-seed, and enclosed in a glass tube, which will continue for months and years to emit penetrating particles producing continuously without cessation most obvious luminous and electrical effects upon distant objects, the particles being so minute that no loss of weight can be detected in the pinch of salt from which they are given off.

The sense of smell is of service to animals—

(1) In avoiding enemies and noxious things.

(2) In tracing and following and discriminating prey or other food.

(3) In recognising members of their own species and individuals of their own herd or troop, and in finding their own young and their own nests.

(4) In seeking individuals of the opposite sex at the breeding season.

It is in connection with the last of these services that we come across some of the most curious observations as to the production and perception of odorous particles. Butterflies and moths and some other insects have olfactory organs in the ends of the antennæ and the "palps" about the mouth. The perfumes of flowers have been developed so as to attract insects by the sense of smell, as their colours have been also developed to attract insects by the eye. The insects serve the flowers by carrying the fertilizing pollen from one flower to another, and thus promoting cross-fertilization among separate individual plants of the same species. But probably concurrently with this has grown up the production of perfume by the scales on the wings of moths and butterflies—perfumes which have the most powerful attraction for the opposite sex of the same species. Curiously enough (for these perfumes might very well exist without being detected by man) some of the perfumes produced by butterflies are "smellable" by man. That of the green-veined white is described as resembling the agreeable odour of the lemon verbena. It is produced by certain scales on the front border of the hinder wings of the male insects, and not at all by the females, who are, however, attracted by it, and flutter around the sweet-smelling male. Other male butterflies produce a scent like that of sweet briar, others like honeysuckle, others like jasmine, and so attract the females. Other butterflies are known which produce repulsive odours, and so protect themselves from being eaten by birds and lizards. Again, there are moths (for instance, the emperor moth, Saturnia), the females of which produce a perfume which attracts the males, and is of far-reaching power. The French entomologist, Fabre, placed one of these female moths in a box covered with net-gauze, and left it in a room with open window, facing the countryside. In less than an hour the room was full of male emperor moths—more than a hundred arrived, although none had been previously visible in the neighbourhood. They crowded over the box, and even afterwards, when the female moth had been removed, the perfume remained in the box, and the male moths eagerly sought it. The perfume must have carried far from the room where the female was, out into the woods where it was perceived, and followed up to its source by the male moths.

Such perfumes are very generally produced by little pockets or glands in the skin, the secretion having, in the case of insects, birds and mammals, an oily nature. In mammals they are largely produced by both males and females, and serve to attract the sexes to one another. Hairs are situated close to the minute odoriferous glands and serve an important part in accumulating and diffusing the characteristic perfume. Musk and civet are of this nature, and it is a significant fact that these substances are used as perfumes by human beings. It would seem as though mankind had lost either the power of satisfactorily perceiving the perfumes naturally produced by the human skin, or that the production of such perfumes had for some reason diminished. Either condition would account for the use by mankind of the perfumes of other animals and of flowers. There are a variety of odorous substances produced by different parts of the human body, of which some are agreeable and others disagreeable. One of the most curious facts in regard to odorous bodies is the close resemblance between agreeable and repulsive odours, and the readiness with which the judgment of human beings may pronounce the same odour agreeable at one period or place, and disagreeable at another. There also seems to be a "dulling" of the power to perceive an odour which is a consequence of constant exposure to that odour. Thus the Chinese say that Europeans all smell unpleasantly, the odour resembling that of sheep, although we do not observe it; whilst Europeans notice and dislike the smell of the negro, a smell of the existence of which he is unaware. The blood of animals, including that of man, has, when freshly shed, a smell peculiar to the species, which has not, however, any resemblance to that of the skin or of the waxy glands of the same animal.

It seems that in regard to the exercise of the sense of smell by man, we must distinguish not only greater from less acuteness and variety of perception, but in the case of this sense-organ, as in regard to the others, we must distinguish "unconscious" from "conscious" sensation. All our movements are guided and determined by sensations to touch and sight, and to some extent, of hearing, of which we are unconscious. A vast amount of our sense-experience comes to us and is recorded without our having consciousness of anything of the kind going on. It is probable that the world of smells in which a dog with a fine olfactive sense lives, produces little or nothing in the dog's mind which is equivalent to our conscious perception of degrees of agreeable and disagreeable odours. The dog is simply attracted and repulsed in this direction and in that by the operation of his olfactive organs, without, so to speak, giving any attention to the sensation which is guiding him or being "aware" of it. No doubt at times, and with special intensities of smell, he is, in his way, conscious of a specific sensation. It is probable that whilst man's general acuteness in perceiving and discriminating smells has dwindled (as has that of the apes) in comparison with what it was in his remote animal ancestry, yet he retains a large inherited capacity of unconscious smell-sense, which most of us are unable to recognise, although it is there, operating in ourselves unknown to us and unobserved. The consciousness of smell-sensations is what we value and talk of. It does not extend to the more primal smell-excitations, except in extraordinary individuals. Thus, it seems to be not improbable that we are attracted or repelled by other human individuals by the unconscious operation upon us of attractive or repulsive odours, and that the unaccountable liking or disliking which we sometimes experience in regard to other individuals is due to perfumes and odours emanating from such persons, which act upon us through our olfactory organs without our being conscious of the fact. It seems that we can thus arrive at a probable explanation of the universality of the habit of kissing, and of "what is that thing we call a kiss." It is not consciously used among civilised populations as a deliberate attempt to smell the person kissed, but it nevertheless serves to allow the unconscious exercise of smell-preference, testing, and selection, with which are mingled, more or less frequently, moments of conscious appreciation of the complex of odours appertaining as an individual quality to the person kissed.

RAY LANKESTER

LAUGHTER

The ancients associated laughter with the New Year. I am not sure whether or no it is of good omen to begin the New Year with laughter. Omens are such tricky things that I have given up paying any attention to them. One would think it might be held to be unlucky to stumble on the doorstep as you set out from home, but the old omen-wizards, apparently from sheer love of contradiction, said, "Not at all! It is unlucky to stumble as you come into the house, and therefore it is lucky to stumble as you go out!"

What is laughter? It is a spasmodic movement of various muscles of the body, beginning with those which half close the eyes and those which draw backwards and upwards the sides of the mouth, and open it so as to expose the teeth, next affecting those of respiration so as to produce short rapidly succeeding expirations accompanied by sound (called "guffaws" when in excess) and then extending to the limbs, causing up and down movement of the half-closed fists and stamping of the feet, and ending in a rolling on the ground and various contortions of the body. Clapping the hands is not part of the laughter "process," but a separate, often involuntary, action which has the calling of attention to oneself as its explanation, just as slapping the ground or a table or one's thigh has. Laughter is spontaneous, that is to say, the movements are not designed or directed by the conscious will. But in mankind, in proportion as individuals are trained in self-control, it is more or less completely under command, and in spite of the most urgent tendency of the automatic mechanism to enter upon the progressive series of movements which we distinguish as (1) smile, (2) broad smile or grin, (3) laugh, (4) loud laughter, (5) paroxysms of uncontrolled laughter, a man or woman can prevent all indication by muscular movement of a desire to laugh or even to smile. Usually laughter is excited by certain pleasurable emotions, and is to be regarded as an "expression" of such emotion just as certain movements and the flow of tears are an "expression" of the painful emotion of grief and physical suffering, and as other movements of the face and limbs are an "expression" of anger, others of "fear." The Greek gods of Olympus enjoyed "inextinguishable laughter."

It is interesting to see how far we can account for the strange movements of laughter as part of the inherited automatic mechanism of man. Why do we laugh? What is the advantage to the individual or the species of "laughing"? Why do we "express" our pleasurable emotion and why in this way? It is said that the outcast diminutive race of Ceylon known as the Veddas never laugh, and it has even been seriously but erroneously stated that the muscles which move the face in laughter are wanting in them. A planter induced some of these people to camp in his "compound," or park, in order to learn something of their habits, language, and beliefs. One day he said to the chief man of the little tribe, "You Veddas never laugh. Why do you never laugh?" The little wild man replied, "It is true; we never laugh. What is there for us to laugh at?"—an answer almost terrible in its pathetic submission to a joyless life. For laughter is primarily, to all races and conditions of men, the accompaniment, the expression of the simple joy of life. It has acquired a variety of relations and significations in the course of the long development of conscious man—but primarily it is an expression of emotion, set going by the experience of the elementary joys of life—the light and heat of the sun, the approach of food, of love of triumph.

Before we look further into the matter it is well to note some exceptional cases of the causation of laughter. The first of these is the excitation of laughter by a purely mechanical "stimulus" or action from the exterior, without any corresponding mental emotion of joy—namely by "tickling," that is to say, by light rubbing or touching of the skin under the arms or at the side of the neck, or on the soles of the feet. Yet a certain readiness to respond is necessary on the part of the person who is "tickled," for, although an unwilling subject may be thus made to laugh, yet there are conditions of mind and of body in which "tickling" produces no response. I do not propose to discuss why it is that "tickling," or gentle friction of the skin produces laughter. It is probably one of those cases in which a mechanism of the living body is set to work, as a machine may be, by directly causing the final movement (say the turning of a wheel), for the production of which a special train of apparatus, to be started by the letting loose of a spring or the turning of a steam-cock, is provided, and in ordinary circumstance is the regular mode in which the working of the mechanism is started. The apparatus of laughter is when due to "tickling" set at work by a short cut to the nerves and related muscles without recourse to the normal emotional steam-cock.

Then we have laughter which is purely due to imitation and suggestion. People laugh because others are laughing, without knowing why. This throws a good deal of light on the significance of laughter. It is essentially a social appeal and response. Only in rare cases do people laugh when they are alone. Under conditions which in the presence of others would cause them to laugh they only "chuckle" or smile, and may, though ready to burst into laughter, not even exhibit its minor expressions when alone. On the other hand, some sane people have the habit of laughing aloud when alone, and there is a recognised form of idiocy which is accompanied by incessant laughter, ceasing only with sleep. Then there is that peculiar condition of laughter which is called "giggling," which is laughter asserting itself in spite of efforts made to restrain it, and frequently only because the occasion is one when the "giggler" is especially anxious not to laugh. This kind of "inverted suggestion," as in the case where an individual "blurts out" the very word or phrase which he is anxious not to use, is obviously not primitive, but connected with the long training and drilling of mankind into approved "behaviour" by "taboos" and restrictive injunctions. Efforts to behave correctly, by causing anxiety and mental disturbance in excitable or so-called "nervous" subjects, lead to an over mastering impulse to do the very thing which must not be done!

It seems that laughter has its origin far back in the animal ancestry of man, and is essentially an expression to others of the joy and exhilaration felt by the laugher. It is an appeal through the eye and ear for sympathy and comradeship in enjoyment. Its use to social animals is in the binding together of the members of a group or society in common feeling and action. Many monkeys laugh, some of them grinning so as to show the teeth, partly opening the mouth and making sounds by spasmodic breathing, identical with those made by man. I have seen and heard the chimpanzees at the Zoological Gardens laugh like children at the approach of their friend and my friend, the distinguished naturalist, Mr. George Boulenger, F.R.S., recognising him among the crowd in front of their cage when he was still far off. And I have often made chimpanzees laugh—"roar with laughter," and roll over in excitement—by tickling them under the arms. The saying of Aristotle (inscribed over the curtain of the Palais Royal Theatre in Paris) that "Laughter is better than tears, because laughter is the speciality of man," is not true. Not only do the higher apes and some of the smaller monkeys laugh, but dogs also laugh, although they do not make sounds whilst indulging in "spasms of laughter." But their distant cousin, the hyena, does laugh aloud, and its laughter agrees with that of the dog and with the laughter of children and grown men in simpler moods in that it is caused by the pleasurable emotion set up by the imminent gratification of a healthy desire. The hyena laughs, the dog grins and bounds, the child laughs and jumps for joy at the approach of something good to eat. But it is a curious fact that the whole attitude is changed when the food is within reach, and the serious business of consuming it has commenced! Nor, indeed, is the satisfaction which is felt after the gratification of appetite accompanied by laughter. It seems that the display of the teeth by drawing back the corners of the mouth, which is called a "grin," and is associated in many dogs with a short, sharp, demonstrative bark, and in mankind with the cackle we call a "laugh," is a retention, a survival, of the playful, good-natured movement of gently biting or pulling a companion with the teeth used by our animal ancestors to draw attention to their joy and to communicate it to others. Gradually it has lost the actual character of a friendly bite; the fore-feet or hand pull instead of the teeth; the sound emitted has become further differentiated from other sounds made by the animal. But the movement for the display of the teeth, though no longer needed as a part of the act of gripping, remains as an understood and universal indication of joy and kindly feeling. So universal is it that this friendly display of the teeth under the name "smile" is attributed to Nature, to Fortune, and to deities by all races of men when those powers seem to favour them.

Laughter is, then, in its essence and origin, a communication or expression to others of the joyous mood of the laugher. There are many and strangely varied occasions when laughter seizes on man, and it is interesting to see how far they can be explained by this conception of the primary and essential nature of the laugh, for many of them seem at first sight remote from it. There is, first of all, the laughter of revivification and escape from death or danger. After railway accidents, earthquakes, and such terrible occurrences, those who have been in great danger often burst into laughter. The nervous balance has been upset by the shock (we call them "shocking accidents"), and the emotional joy of escape, the joy of recovered life, asserts itself in what appears to the onlooker to be an unseemly, an unfeeling laugh. It is recorded that one of the entombed French coal miners, who two years ago were imprisoned without food or light for twenty days a thousand feet below in the bowels of the earth, burst into a ghastly laugh when he was rescued and brought to the upper air once more. The Greeks and Romans in some of their festal ceremonies made the priest or actor who represented dead nature returning to life in the spring, burst into a laugh—a ceremonial or "ritual" laugh. Our poets speak of the smiles, and even of the laughter of spring, and that is why laughter is appropriate to New Year's Day. It is the laughter of escape from the death of winter and of return to life, for the true and old-established New Year's Day was not in mid-winter, but a quarter of a year later, when buds and flowers are bursting into life. It is recorded by ancient writers that the "ritual laugh" was enforced by the Sardinians and others who habitually killed their old people (their parents) upon their victims. They smiled and laughed as part of the ceremony, the executioners also smiling. The old people were supposed to laugh with joy at the revivification which was in store for them in a future state. So, too, the Hindoo widows used to laugh when seated on the funeral pyre ready to be burnt. So, too, is explained (by Reinach) the laughter of Joan of Arc when she made her abjuration in front of the faggots which were to burn her to death. Her laugh was caused by the thought of her escape from persecution and of the joyful resurrection soon to come. It was not an indication that she was not serious, and that her abjuration of witchcraft was a farce, as her enemies asserted.

More difficult to explain is the laughter excited by scenes or narrations which we call ludicrous, funny, grotesque, comic; and still more so the derisive and contemptuous laugh. Caricature or burlesque of well known men is a favourite method of producing laughter among savages as well as civilised peoples. Why do we laugh when a man on the stage searches everywhere for his hat, which is all the time on his head? Why do we laugh when a pompous gentleman slips on a piece of orange-peel and falls to the ground, or when one buffoon unexpectedly hits another on the head, and, before he has time to recover, with equal unexpectedness hooks his legs with a stick and brings him heavily to the ground? Why did we laugh at the adventures of Mr. Penley in "Charley's Aunt"? In all of these "ludicrous" affairs there is an element of surprise, a slight shock which puts us off our mental balance, and the subsequent laughter, when we realise either that no serious harm has been done or that the whole thing is make-believe, seems to partake of the character of the "laugh of escape." It is caused by a sense of relief when we recognise that the disaster is not real. We laugh at the "unreal" when we should be filled with horror and grief were we assured that there was real pain and cruelty going on in front of us. The laughter caused by grotesque mimicry or caricature of pompous or solemn individuals seems to arise from the same (more or less unconscious) working of the mind as that caused by some unexpected neglect of those social "taboos" or laws of behaviour which we call modesty, decency, and propriety. They either cause indignation and resentment in the onlooker at the neglect of respect for the taboo, or, on the contrary, the natural man, long oppressed by pomposity or by the fetters of propriety imposed by society, suddenly feels a joyous sense of escape from his bonds, and bursts into laughter—the laughter of a return to vitality and nature—which is enormously encouraged and developed into "roars of merriment" by the sympathy of others around him who are experiencing the same emotion and expressing it in the same way.

The laugh of derision and contempt and the laugh of exultation and triumph are of a different character. I cannot now discuss them further than to say that they are either genuine or pretended assertions of joy in one's own superior vitality or other superiority. The "sardonic smile" and "sardonic laughter" have been supposed by some learned men to refer to the smiles of the ancient Sardinians when stoning their aged parents. But they have no more to do with Sardinians than they have with sardines or sardonyx. The word "sardonic" is related to a Greek word which means "to snarl," and a sardonic grin is merely a snarl. In it the teeth are shown with malicious intent, and not as they are in the benevolent appeal of true laughter. Mrs. Grote, the wife of the great historian (who was herself declared by a French wit to furnish the explanation of the word "grotesque"), wrote of "Owen's sugar-of-lead smile"—referring to the great naturalist, Richard Owen. There was no malice in the description, for he had, as some others have, a very sweet smile, accompanied by a strangely grave and disapproving glare in his large blue prominent eyes. It was only apparently sugar of lead; really, it was sugar of milk—the milk of human kindness. The smile of the lost picture called "La Gioconda" is by fanciful people regarded as something very wonderful. It is really the clever portraiture of the habitual "leer" of a somewhat wearied sensual woman. It had a fascination for the great Leonardo, but no profound significance.

FATHERLESS FROGS

One of the most interesting discoveries of recent date in regard to the processes which go on in that all-important material—protoplasm—which is the physical basis of life and the essential constituent of "cells"—those minute corpuscles of which all living bodies are built—was made in 1910 by a French naturalist, M. Bataillon, and has been examined and confirmed by another French biologist, M. Henneguy. To explain this discovery, a few words as to well-known facts are necessary. It is well known that if we isolate a female frog at the egg-laying season and let her swim in perfectly pure filtered water, and proceed to deposit some of her eggs in that water, the eggs will not germinate; they remain unchanged for a time and then decompose—become, in fact, "rotten." It is a matter of common knowledge that it is necessary for the eggs to be "fertilised" in order that they may start on that series of changes and growth which we call "development," and become tadpoles and eventually young frogs. The "fertilisation" of the frog's eggs is effected in ordinary conditions by the presence in the water of the pond, into which the female sheds them, of microscopic sperm-filaments (often called spermatozoa, or simply "sperms") which are shed into the water at the same time by the male frog.

The egg (the blackish-brown spherical body, as big as a rape-seed, which is imbedded in a thin jelly, and is familiar to those who are drawn by curiosity to look into the waters of wayside ponds in spring) is a single cell or corpuscle of protoplasm distended with dark-coloured and other granules of nutrient substance. A single sperm (though requiring the microscope to render it visible) is also a single cell. It is a minute oval body, with a long serpentine tail of actively undulating protoplasm. Hundreds of thousands of these are shed into the water at the breeding season by the male frog. One is enough to fertilise the egg. The sperm-cells swim in the water, and are chemically attracted by the eggs. As there are so many sperms, one of them is sure to reach each black egg-sphere. It drives its way into the substance of the egg, making a minute hole in its surface; then the protoplasm of the sperm fuses with the protoplasm of the egg, and becomes intimately mixed with it. The egg-cell has a "nucleus," that dense, peculiar, deep-lying, and well-marked "kernel" of its protoplasm which all cells have. It is of essential importance in the life and activity of the cell. The sperm-cell has also a "nucleus," and now (as has been carefully ascertained) the nucleus of the sperm and the nucleus of the egg-cell unite and form one single nucleus. The egg is thereupon said to be "fertilised"—that is to say, "rendered fertile." It at once commences to move. Its surface ripples and contracts and nips in deeply, so that the sphere is marked out into two hemispheres. These are two "cells," or masses of protoplasm, adhering to each other. Each is provided with its own distinct nucleus or cell-kernel, for the first step in the division of the egg-sphere is the division within it of its newly constituted nucleus into two, each half consisting of nearly equal proportions of the mingled substance of the sperm-nucleus and the egg-nucleus. The two first cells or hemispheres again divide, and so the process goes on until the little black egg has the appearance of a mulberry, each granule of the berry being a cell provided with its own nucleus derived from the original nucleus formed by the fusion of the nuclei of the paternal and maternal cells. In the course of a day or two the division has proceeded so far that the resulting "cells" are so small as to be invisible with a hand-glass, and require one to use a high magnifying power in order to distinguish them. And there are hundreds of them; the whole mass of the "egg" within, as well as on the surface, has divided into separate cells. They go on multiplying, take up water, and nourish themselves on the granular nutritive matter present from the first in the egg-cell. The little mass elongates, increases in size, and gradually assumes the form of a young tadpole.

We see, then that the process of fertilisation consists in two things, the latter of which necessitates the former, viz. in the breaking or penetration of the surface of the egg-cell by the active sperm filament and second in the fusion of the substance of the sperm filament with that of the egg in such a way that there is a distinct and intimate fusion of the nucleus of the sperm filament with the nucleus of the egg-cell. The recent discovery of M. Bataillon is this, viz. that you can make the frog's egg develop in a perfectly regular way and become a tadpole and then a young frog without the admission to it of a sperm-filament or of any substance derived from the male frog. All you have to do—and the operation, though it sounds easy and simple, is an exceedingly delicate and difficult one—is to prick with a fine needle the surface of the little black egg-sphere (not merely of the jelly surrounding it) when it is shed by the female frog into perfectly pure water free from sperms or anything of the sort. The slight artificial puncture acts as does the natural puncture by the swimming sperm-filament, and is sufficient! The egg proceeds to develop quite regularly. There is no fusion of the nucleus of the egg-cell with any matter from the outside; no paternal "material" is introduced, but the nucleus of the egg-cell divides just as though there had been! The whole progeny of cells, successively formed, are the pure offspring of the maternal egg-cell and its nucleus. The tadpoles and young frogs so produced are examples of what is called "parthenogenesis"—that is to say, virginal reproduction—reproduction without fertilisation by material derived from a male parent! The needle, which gives off no material, but simply makes a tiny break in the surface of the egg, does all that is necessary!

To those not acquainted with all that has been ascertained as to the reproduction of lower animals such as insects, crustaceans, and worms, this discovery will appear more astonishing than it really is. We know of many lower animals in which the egg-cells produced by the females do regularly and naturally develop without the intervention of a male and without fertilisation. In an earlier volume[7] of this "Easy Chair Series" I wrote of this curious subject, and described the virgin reproduction or parthenogenesis of the hop-louse and other plant lice, of some moths, of some fresh-water shrimps, and of the queen bee (who produces only drones by eggs which are not fertilised). But I had to point out then that no case was known of "parthenogenesis"—that is to say, reproduction by unfertilised eggs—among the whole series of vertebrate animals, the fishes, amphibians, reptiles, birds, and mammals. The chief point of novelty in M. Bataillon's discovery is that we have now an experimental demonstration of parthenogenesis in a vertebrate animal, and in one so highly organised as the frog. And equally interesting, indeed more important from the point of view as to the real meaning and nature of fertilisation, is the mode in which the parthenogenesis of the frog is set going, namely, by a mere prick of the surface film of the ripe egg!

There have, however, been important experiments on the subject of the development of eggs without fertilisation in recent years, prior to these discoveries as to the frog's egg. A favourite subject for such inquiries is the sea urchin (Echinus of different kinds). The female sea urchin, or sea egg, like its close allies the star fishes, lays a great number of very transparent minute eggs (each about the 1/200th of an inch in diameter) in sea-water, and they are there fertilised by the mobile sperm filaments discharged by the males. The eggs are so transparent and so easily kept alive in jars of sea-water that there is no difficulty in watching under the microscope the penetration of the egg by a sperm, and the fusion and other changes in the nuclei. Delages of Paris, and Loeb of California, have made valuable studies on these eggs. Loeb has shown that they may be artificially started on the course of development and cell division without fertilisation—simply by the action of minute quantities of simple chemicals (fatty acids, etc.) introduced into the sea-water by the experimenter. These chemicals appear to act on the delicate pellicle which forms the surface of the egg-cell in much the same way as the prick of a needle acts on a frog's egg. A limited and delicately adjusted disturbance of the cohesion (or of the surface-tension) of the egg-cell seems to be all that is necessary for starting the egg-cell on its career of development. It becomes, in the light of these experiments, not so much a wonder that egg-cells should develop "on their own," but that they do not more frequently do so. It must be remembered that the "germination" and development of unfertilised eggs, even when the whole range of animals and plants is taken into account (for plants also are reproduced by single cells identical in character with the egg-cells and sperm-cells of animals), that is to say, the existence of "parthenogenesis" as a natural, regularly recurring process, is exceptional. We must distinguish cases in which it regularly occurs as part of the life-history of an animal or plant from cases in which it has been successfully brought about by experimental "artificial" methods designed by man. The plant-lice "naturally" reproduce through the summer by unfertilised eggs producing only females, but in the first cold of autumn males are hatched from some of the eggs, and the eggs of this generation are fertilised and bide through the winter, hatching in the following spring. Some few moths and flies also reproduce naturally during summer by unfertilised eggs, and the brine-shrimps and some other fresh-water shrimps produce "fatherless" broods from their eggs, sometimes for years in succession, until "one fine day" some males are hatched, owing to what causes we do not know. The queen bee naturally and regularly lays a certain number of unfertilised eggs, and these produce, not females as do the unfertilised eggs of plant-lice, etc., but male bees—the drones—and it is only from such eggs that the drones of bees are born. These are the chief cases of regular and natural parthenogenesis, but there are others which might be enumerated.

On the other hand, examples of artificially induced development of eggs, not fertilised, are very few. The first known came accidentally to notice. Female silkworm moths reared in confinement sometimes lay eggs when kept apart from the male, and these have been found to hatch, and give rise to caterpillars, which were not reared to maturity. Other moths bred by collectors behaved in the same way, but the grubs were reared to maturity, and three successive generations of "fatherless" moths were obtained. In these cases the hatching of unfertilised eggs is not known to occur in a state of nature, although it probably occurs occasionally. It has also been observed—an important fact when considered with the history of the frog's egg and the needle—that "brushing" the unfertilised eggs of the silkworm and other moths, that is to say, gently polishing the little egg-shells with a soft camel's-hair brush, has the effect of starting development. Taking two lots of unfertilised eggs adhering to slips of paper, as laid by the mother moth, it is found that those gently brushed will hatch, whilst those not brushed will either not hatch at all, or in very small number. The brushing seems to disturb the equilibrium of the protoplasmic egg-cell within the egg-shell just sufficiently to set it going—going on its course of division and development. The only other case of "artificially-induced parthenogenesis" at present recorded is that of the common frog, due to M. Bataillon. There are questions of great interest still to be made out as the result of his discovery. Can the fatherless brood be reared to maturity and again made to yield a fatherless generation? What is the precise structure of the nuclei of the cells which originate from the nucleus of the egg-cell only, and not from a nucleus formed by the fusion of that with a sperm-cell nucleus? These and similar questions are the motive of further careful study now in progress.

The important conclusion is forced upon us by these experiments with a needle, that even in so typical and highly organised a creature as one of the higher or five-fingered, air-breathing vertebrates, the egg-cell does not require any material admixture from the sperm-cell in order that it may successfully germinate and develop, but only a disturbance of equilibrium, which can be administered as well by a needle's point as by a sperm-filament! Yet the whole process of sexual reproduction undoubtedly has, as its origin and explanation, the fusion in the first cell of the new generation from which all the rest will arise, of the material of two distinct individuals. Thus the qualities of the young are not a repetition of the qualities of one parent, nor are they a mere mixture of the qualities of both parents (for contradictory qualities cannot mix). They are a new grouping of qualities comprising some of the one parent and some of the other and hence a great opportunity for variation, for departure from either parent's exact "make-up," is afforded, and for the selection and survival of the new combination. It is, it would seem, only in exceptional cases and for limited periods that uni-sexual or fatherless reproduction can be advantageous to a species of plant or animal. Such cases are those in which abundant food, present for a limited season, renders the most rapid multiplication of individuals an advantage to the species. But after this exceptional abundance has come to an end, the more usual process of reproduction by fertilised eggs (also necessary and advantageous for the preservation of the race by "natural selection in the struggle for existence" of the new varieties so produced) is resumed until again the abundant food is present, as in the annual history of plant lice and the plants on which they feed.

[7] "Science from an Easy Chair," Methuen and Co., 1910.

RAY LANKESTER

PRIMITIVE BELIEFS ABOUT FATHERLESS PROGENY

In the preceding chapter I related the curious and exceptional cases of "fatherless reproduction" by means of true egg-cells, those cells of special nature produced in the organs called "ovaries," present in all but the simplest animals and plants. These egg-cells are usually, with elaborate sureness and precise mechanism after liberation from the ovary, fertilised by (that is to say, fused with) the complemental reproductive cells—the sperm-filaments—produced by other individuals, the males.

But we must not forget—and, indeed, one should not enter on the consideration of this subject without a knowledge of the fact—that vast numbers of animals and plants reproduce themselves "asexually," as it is termed, namely, by breaking-off or separating buds, branches, or other good solid bits of their structure which, when thus separated, are capable of individual life and growth. Thus plants very largely multiply, using this method in addition to the sexual method of egg-cells and sperm-cells. One may take "cuttings" from plants and rear them, and plants also "cut" or detach such bits themselves, in the form of runners, of dividing bulbs, of bulbules, and such reproductive growths seen on the lily, on the viviparous, alpine grass, and many other plants. Even a bit cut off from the leaf of a plant (for instance, a begonia) will sprout, root itself, and grow into a completely formed and healthy individual. Animals, too, such as polyps or zoophytes, and many beautiful and elaborate worms, multiply by "fission," dividing into two or more parts, each of which becomes a complete animal. This process is not seen in any fish, amphibian, reptile, bird, or mammal, nor in molluscs, nor in insects, crustaceans, myriapods, and arachnids (spiders and scorpions). It is almost wholly confined to lower animals (worms and polyps) and to plants, and hence is often called "vegetative reproduction." The most remarkable case of its appearance among higher forms is that of the marine Ascidians, or tunicates—close allies of the true vertebrates—where reproduction by budding and the formation of wonderfully elaborate star-like forms produced by budding and the cohesion of the budded individuals as one composite individual are well known. Their beautiful shapes and colours have been reproduced in hundreds of exquisite pictures by our great artist-naturalists. We thus have to recognise that there are two distinct kinds of reproduction in living things. One is "asexual," by means of division or separation of large or special masses of their existence, made up of ordinary tissue cells. Co-existing with this, often in the same individuals, is the other method, the "sexual," by means of detached egg-cells and sperm-cells which are thrown off from the parents, and do not (except in rare instances) proceed to develop unless the egg-cell is "fertilised" by the fusion with it of a sperm-cell.

The whole subject of the reproduction of animals and plants was, until the introduction of the microscope, involved in obscurity and mystery. The Greeks and Romans had necessarily very imperfect and erroneous notions on the subject, and it was not until 300 years ago that William Harvey, the discoverer of the circulation of the blood, declared, as a general law, that every living thing is born from an egg. During that 300 years his conclusion has been examined and modified, corrected and expanded, and the microscope has at last enabled us to see and follow the excessively minute particles and structures by which sexual reproduction is effected. Harvey's dictum was a step in advance when it was made, for previously the belief was current that living things were "bred" in all sorts of queer ways. It was supposed that the putrefying flesh of a dead animal actually was converted by a sudden process into maggots, and that rotten wood would breed, out of its own substance, ships' barnacles and even young geese and mice—an opinion contested only 200 years ago by Sir Thomas Browne! No difficulty was felt in admitting that whole swarms of insects, fishes, and even herds of larger beasts were spontaneously generated from mud, from putrid matter, or from the waters of the sea. That, indeed, was the popular notion set forth by the poet, John Milton, as to the mode in which living things were "miraculously" brought into existence at the beginning of things by the "fiat" of the Creator. What more probable than that such a creation should still be, here and there, at work? However, not three centuries ago, actual experiment gradually convinced the learned that maggots are bred in a dead body only from the eggs laid by parent flies, as shown by the Italian Redi in 1668 who found that no maggots were bred when he simply excluded the flies from access to the dead body by covering it with wire gauze, but that the blow-flies swarmed on the gauze and vainly laid their eggs on it! It was only gradually recognised that birth by means of eggs or germs extruded from parental organisms of the same history and character as their offspring is the explanation of all such swarms of flies, worms, and even mushrooms and moulds as had been formerly ascribed to a mysterious power of breeding these organisms possessed by inanimate dirt and refuse.

In spite of this progress in knowledge the belief in "spontaneous generation" of such excessively minute organisms as the bacteria and yeasts was general until Theodore Schwann in 1836 performed with them just the same experiment as Redi had performed with blow-flies in 1668. He showed that if a putrescible liquid (for instance, soup) were boiled in a retort so as to destroy all germs, and then the open neck of the retort was kept heated in a flame, so that no floating germs could enter alive, the soup did not putrefy, and no bacteria or other organisms appeared in it. The old notions, nevertheless, survive to this day. Peasants, fisher-folk, and even uneducated wealthy countrymen cling to them with the confidence arising from profound ignorance. And occasionally a man of some scientific training and knowledge astonishes the world by a futile attempt to show that the old fancies were true in regard, at any rate, to the lowest microscopic forms of life. But these are but the echoes of the past; we do not believe nowadays in "spontaneous generation," nor in sudden transformations of lower into higher forms of life. The doctrine, "omne vivum e vivo"—every living thing (in the present condition of our earth) is born from a living thing—is now held by scientific investigators as a reasonable generalisation of experience.

On the other hand, Harvey's dictum, "Every living thing comes from an egg," is only true in a limited sense, namely, that whilst the individual among most larger animals and plants is always traceable to an egg-cell detached from a parental individual of a like kind of species, there are whole groups and series of lower animals and most plants in which the individual born or "developed" from an egg-cell does not proceed when grown to full size to reproduce in turn by eggs and fertilising sperms, but divides into two or more individuals or gives off detached buds or reproductive bulbs, which become separate individuals, and only after these and several successive generations of individuals have been thus produced "asexually," by fission or by budding, does a generation appear which produces true egg-cells and sperm-cells and reproduces by their means. Thus it is true that the individuals "budded off" or separated by fission from an asexual parent can be ultimately traced through one or more generations of previous asexual parents to an egg-cell produced and fertilised in the regular way, and with this important modification Harvey's dictum is justified. These facts and the wonderful histories of the animals and plants in which egg-and-sperm-producing generations "alternate" with generations which multiply by fission and budding have only been worked out in detail and by the aid of the microscope during the great century of scientific discovery which lies just behind us. Often the two generations, reproducing, the one by fission, the other by egg and sperm-cells, are alike in appearance, but often they are very different, and have naturally been supposed at first to have nothing to do with each other.

Thus some of the little "coralline polyps" and other most beautiful little marine flower-like polyps attached to rocks, weeds, and shells in the sea reproduce by budding and division. But after a period of such growth and such budding they produce on their stalks—jelly-fish! These jelly-fish are budded and thrown off by them, as glass-like swimming bells, which lead an independent life, seize prey, nourish themselves, and grow to a size varying from that of a sixpence to that of a cart-wheel. These "bells" are commonly known as "jelly-fish." They discharge thousands of egg-cells into the sea and fertilise them with sperms! From those fertilised eggs grow young polyps, which fix themselves to rocks or weeds, and grow up to bud and multiply by fission, and eventually to produce again by fission a generation of jelly-fishes! Such a marvellous history of alternating modes of reproduction has been discovered, and described in greatest microscopic detail and with most ample pictorial representations of all the minutest structures of the organisms studied, not only in many marine polyps, but also in the case of many parasitic worms, such as the tape worms and the liver-flukes. Some of the most fascinating cases, on account of the beauty of the little creatures concerned, are found amongst the surface-swimming Ascidians of the sea—the glass-like Salps. But our common ferns and mosses also show this same alternation of sexual and sexless generations, the two generations differing greatly in size, form, and structure from one another, whilst the whole story of "flowers" and their structure is bound up with a wonderful "telescoping" or rolling of the two generations (sexless and sexual) into one plant!

It was not until long after Harvey's time that these things were understood, and there was every excuse—in the absence of observation of the facts, especially those yet to be revealed by the microscope—for the erroneous suppositions and explanations which were formerly entertained as to the mode of reproduction of the less familiar plants and animals. If we go back to the starting-point of European science, to the great Aristotle, we find that he had formed singularly correct conclusions as to the reproduction of the larger kinds of animals, though he knew nothing about "sperms," having no microscope, and only regarded the fluid produced by male animals as exercising a fertilising effect on the eggs, which in many instances are large enough for anyone to see. But, of course, he could not have any knowledge of the egg-cell, nor does he say anything about the reproduction of plants. Later, however, the sexuality of flowering plants was taught by his pupils, and at the time of the Roman Empire there was a very definite belief among learned men (such as Pliny) that the larger plants and animals reproduce by eggs or by seeds produced by the females which require to be "fertilised" by a product formed in the males—the spermatic fluid in the case of animals and by the pollen in the case of a few flowering plants (e.g. the date-palm). But there was no idea of holding this as a general and universal law. From Pliny to Harvey and later, those who concerned themselves with natural history accepted without difficulty any strange accounts or appearances as to the reproduction or the sudden production in fanciful and astonishing ways of the lower and smaller animals and plants. They did not expect these inferior creatures to have the same methods of reproduction as the higher and bigger creatures. It is only now, since the later years of the nineteenth century, that we are able to show that all animals and plants, even the minutest microscopic kinds, reproduce by the formation and separation of egg-cells, and that these egg-cells are (in all but a few exceptional cases) fertilised by sperm cells, which are smaller than the egg-cells, and usually provided with active swimming filaments.

Not only did our mediæval ancestors believe all sorts of fancies as to the propagation of lower animals and plants, but they were quite prepared to accept stories as to reproduction in the case of higher animals, and even in mankind, by irregular methods, such as parthenogenesis, or the defect of an ordinary male parent. In the Middle Ages in Europe, and earlier in the East, the belief in the frequent occurrence of the birth of a child which had no human male parent was common. It was, so to speak, an admitted though irregular occurrence. A very curious thing is that when such cases were supposed to occur, they were not ascribed to any natural process such as we now recognise in the "parthenogenesis" of insects and crustaceans, but to the visitation of the mother by a spirit—a floating, volatile demon or angel (known as an "incubus" in the Middle Ages) beneficent or malicious as the case might be. Stories of the nocturnal visits of these mysterious ghostly "incubi" are on record in great number and variety, both in European and Oriental tradition and legend. There seems to have been a readiness to believe the theory of paternity from among the hidden world of goblins, fairies, and sprites which was very naturally made use of by a woman and her relatives when she could not produce the father of her child.

We come across examples of such beliefs in invisible agents of paternity even among the more cultivated Romans. Thus Virgil in his "Georgics" cites as a fact that mares are fertilised by the wind. His words are given on the next page.

It is now known that, quite apart from any motive of concealment of the true paternity of their offspring, some of the native tribes of Australia have the belief that, as the regular and normal thing, children are begotten by strange fairy-like spirits which haunt the rocks and trees of certain localities and enter the future mother as she passes by these haunted rocks and trees.

These Australian "black fellows" hold that the human father counts for nothing in the matter. The belief of these Australian savages is referred to by writers on the subject (Mr. Andrew Lang and others) as "the spiritual theory of conception." There are some reasons for thinking that this curious theory and the accompanying ignorance as to the natural causes of conception were widely spread among primeval men. The fact that most trees are fertilised by the wind (which carries to their female flowers the invisible powder, or pollen, of the male flowers, conveyed in the case of smaller plants which have gay-coloured flowers by bees and butterflies) may have been noticed by primitive man, and have started the belief that there are fertilising spirits or demons in the air. However the fancy arose, it is only a parallel to the strange fancies as to spontaneous generation of all sorts of animals and plants current 200 years ago among civilised men. And, further, it is worth noting that the uncanny belief in the "incubus" which was generally prevalent in the Middle Ages may possibly be considered as a survival in (or incursion into) Europe of the primitive spiritual theory of all human conception, and of the fertilising activity of the haunting spirits of the air which was held by primeval man, and is still found in full force among the Arunta tribes of Australia.

"Ore omnes versæ in Zephyrum stant rupibus altis
Exceptantque leves auras et sæpe sine ullis
Conjugiis vento gravidæ, mirabile dictu."
Georgic iii. 275.
(Facing the west on lofty rocks
All stand and sniff the buoyant breeze
And often—marvellous to tell—
Without conjunction with a sire,
Bear young engendered by the wind.)

THE PYGMY RACES OF MEN

The tradition of the existence of dwarfs, not as isolated examples, but as a race with their own customs, government, and language is familiar among civilised people, and exists among scattered and remote savages. We have all heard of them in that treasury of primitive beliefs—the nursery. Therefore, the fact that there are at this moment in various parts of the world dwarf or pygmy tribes of men, living in proximity to but apart from those races which have a stature identical with our own, has a great fascination and interest. Some few races of men have an average height of an inch, or thereabouts, greater than that of the people of the British Islands, whilst some are shorter by as much as two or three inches. But, on the whole, it may be said that, putting aside the pygmy races, of which I am about to write, mankind generally does not show a very striking range of normal stature—the mass in any race or region of the globe varying from 5 ft. 4 in. to 5 ft. 8 in., and tending to the higher rather than the lower figure.

The pygmy races are sharply separated from normal mankind by as much as a foot, and even more, in average stature, ranging from 4 ft. to something less than 4 ft. 11 in. in height. They are, enumerating them in the order of their purity of race and completeness of their isolation: (1) The Mincopies, or Andaman Islanders; (2) the Congo pygmies (comprising the tribes known as the Akkas, or Tiki-Tikis, the Bambutis, the Watwas, the Obongos, and Bayagas); (3) the bushmen of South Africa; (4) the Aetas of the Philippine Islands; (5) the Samangs of Malacca, and very similar isolated pygmy tribes which have been observed in New Guinea, and also in the Solomon Islands and in Formosa. The Veddas of Ceylon, the Senois of Malacca, and the Toalas of Celebes are apparently races which have resulted from the "crossing" of true pygmies with other normal-statured races inhabiting the islands in which they are found. The Brahouis of

Beloochistan and the "monkey-men," or Bandra-Loks, east of the Indus, appear also to belong to the pygmy race.

Next to their agreement in small size, the most interesting facts about the pygmies we have just enumerated is that, notwithstanding the wide area over which they are found in scattered, isolated communities—viz. from the Congo to South Africa on the one hand, and, on the other hand, from Central Africa to the Indian Ocean, and on to New Guinea, the Philippine Islands, and Formosa—yet they all have short, round skulls of full average brain capacity, and have their hair growing in tightly curled-up peppercorn-like tufts—two characters found combined in no other race. They usually have finely-developed, straight foreheads, and the jaws do not project strongly; the lips are usually fine and thin, and the nose, though very broad, is not always greatly flattened. They are well-shaped, well-proportioned little people, neither grotesque nor deformed. To a great extent their corporeal features suggest an infantile or child-like stage of development, and the same is true of their intellectual condition and of their productions. Their habitations are very primitive, either caves or low clay-made huts, of the shape of half an egg. They do not make pottery, and neither keep herds nor till the ground, contenting themselves with such food as wild fruits and roots and the animals they kill with spear or arrow or capture in traps. They do not mutilate or bedaub their bodies (though the Andamanese indulge in a kind of "tattooing"). Among them the struggle for life does not exist in its more brutal forms. They take care of the sick and feeble, the children, and the old people. Cannibalism is unknown amongst them; they punish murder and theft. They are honest, and, moreover, are monogamous, and punish adultery, which is rare among them. Their religion is remarkably simple. It is limited to reverence for a Supreme Being, without any offering of sacrifice, and they do not worship ancestors nor exhibit the superstitions known as "animism." It has been argued that these characteristics, taken together, indicate a primitive condition of humanity. On the other hand, many writers regard them as degenerate offshoots of negro-like races of larger stature and more complicated mental development.

There is no name by which the whole series of these small-sized people is indicated excepting the ancient designation of "pygmies." Many careful students of human races separate the pygmies of Africa as "negrilloes" from the pygmies of Asia, whom they designate "negritoes," and it is held that the negrilloes (Congo pygmies and bushmen) hold the same relation to African negroes and Zulus as the negritoes (Andamanese, and scattered tribes in New Guinea, the Philippines, Formosa and the Solomon Islands, as well as in Malacca and Annam and in the north-west and in other parts of Hindustan) hold to the full-sized, frizzly haired Papuans. This, no doubt, is a convenient way of stating the case, but the important fact remains that the pygmies of purest race, both of Africa and Asia, have the remarkable characteristics in common which we have noted above. Their bodily and mental peculiarities certainly suggest, whether the suggestion can be verified or not, the former existence in the tropical regions of Africa and Asia of a widely spread pygmy race of uniform character, a race which has been, to a large extent, destroyed by other races of larger and more powerful individuals, but has also in many regions (especially on the Asiatic Continent) intermarried with the surrounding larger people, and given rise to hybrid races. At the same time, it seems that in other regions this race has, by isolation in forests and mountain ranges and by the exercise of special skill in the use of poisoned arrows and in the arts of concealment, evasion, and terrorising, succeeded in maintaining its existence and primitive independence dating from remote prehistoric times.

Whether we regard the pygmies as one race or as the result of local modification of larger races, it is noteworthy that they are of lighter tint than the black races close to or among whom they live. Some, both of the African and Asiatic pygmies, are very dark brown—practically black—but many are of a paler and yellowish tint. We must not forget that the babies and quite young children of negroes are nearly "white." The Asiatic pygmies, notably the Andamanese, are darker than their African fellows. It must necessarily be difficult in studying such a race to make due allowance not merely for admixture of blood from surrounding populations, but to estimate correctly what the little people have learnt in the way of art and habit from their neighbours and what is their own. The Andaman Islanders, though provided with metal by trading, still use the sharp-edged splinters of volcanic glass-stone to shave their heads, which they keep entirely bald!

It is one of the merits of the showman's enterprise in modern times that he brings to a great city like London groups of interesting savages, without imposture and without ill-treatment, and enables us to see and talk with them almost as though we had travelled to their remote native forests. It would certainly be a successful and worthy enterprise on the part of the Anthropological Society of London to start a garden and houses such as those maintained by the Zoological Society, but arranged so as to receive some five or six groups of interesting "savages." The society would be responsible for careful and humane treatment of their guests, and return them after a sojourn, say, of a couple years, to their native country and replace them by specimens of other races. Under the auspices of showmen I have seen Zulu Kaffirs, Guiana Indians, North American Indians, Kalmuck Tartars, South African bushmen, and Congo pygmies in London, besides many hundreds of African negroes of various tribes. Farini's bushmen and Harrison's Congo pygmies were perfect samples of the dwarf race about which I am writing. But I also saw and examined carefully, in 1872, at Naples, with my friend Professor Panceri, the two African pygmies, Tebo and Chairallah, who were the first to reach Europe. They were subsequently adopted by and lived for some years under the care of Count Miniscalchi Erizzo. They were very intelligent, and learnt to read and to write well, and to play difficult music on the piano, with feeling and appreciation. We were especially concerned to determine by the stage of growth of their teeth and other indications whether they were merely ordinary young negroes, as some anthropologists supposed, or really representatives of the dwarf race as asserted by the traveller Miani, who bought them, in exchange for a dog and a calf, in the country of the Mombootoos, south of the Welle River, and west of the Albert Nyanza. They were still young and growing when we examined them, but Tebo ceased growth when he had reached a stature of 4 ft. 8 in. We had no difficulty in coming to the conclusion that they were, when we saw them, really of exceptionally small stature for their age as indicated by the teeth which were in place in their jaws.

Fig. 23.—Copy of a figure from a group drawn on a Greek vase (dating from 300 b.c.), representing a number of the pygmies of the remote Upper Nile engaged in battle. The resemblance of the peaked cap and of the beard to those of the little figures carved by Black Forest peasants and intended to represent the mythical "gnomes" or dwarf mining-elves is noteworthy. (From Saglio and Derenberg's "Dictionnaire des Antiquités Grecs et Romaines.")

The Akkas living near the sources of the Nile were known to the ancient Egyptians, and were the foundation of stories and fabulous exaggerations among the ancient Greeks. Even before Homer these stories existed, and the little people were called "pygmies," which means "of the length of the forearm" (Greek, pugmé). Homer refers to the wars of these pygmies with the cranes, and as a matter of fact the African pygmies do wage a kind of war upon the great cranes which swarm in the marsh-land of their country. Naturally enough the really small size of the African pygmies (they are about 4 ft. in height, some two or three inches less, some as much as eight inches more) was exaggerated by report and tradition, just as the really big eggs of the great extinct ostrich-like bird of Madagascar were represented in the story of Sindbad, in the "Arabian Nights," as being as large as the dome of a temple, and the bird large in proportion. The Egyptians, as we have seen, knew the pygmy Akkas, and Egyptian fact was ever the romance of the Greeks.

Herodotus mentions the African pygmies from beyond the Libyan desert, citing, as is his wont, the accounts of certain travellers with whom he had conversed, and a later Greek writer tells of a pygmy race in India, a statement which our present knowledge confirms. It is a curious fact that Swift's Lilliputians are thus traceable to the Central African dwarf race, for Greek legend related that Hercules visited the country of the pygmies, where on waking from sleep he found one division of the army guarding his right leg, another his left, and others his arms. Hercules got up, swept them all into the lion's skin which he used as a cloak, and went on his way, shaking out his small tormentors from their prison as though they were so many ants. It seems fairly certain that Swift derived the initial scene in his story of Gulliver's adventures among the Lilliputians from this legend.

Miani's pygmies were members of a tribe discovered by the distinguished traveller Schweinfurth, who, in 1870, was the first to visit the country of the Niam-Niam, to the west of the sources of the Nile, and had the honour of showing that the myths of the ancient Greeks as to a nation of pygmies were based on fact, and that the definite words of Aristotle as to the existence of these pygmy people on the upper reaches of the Nile were correct. Schweinfurth found to the south of the Niam-Niam country a tribe of full-statured negroes called the Mombootoos, whose chief, Moonza, kept close to the Royal residence a colony of pygmies who were called in that country by the name "Akkas." Schweinfurth ascertained that they are spread to the number of many thousands along the borders of the great Congo forest and form numerous tribes. They are very generally well treated by their more powerful neighbours, as by Moonza. Partly from fear of their poisoned arrows and their crafty methods of attack and subsequent disappearance into the forest, partly on account of a superstitious dread of them, the Congo pygmies are not only tolerated, but protected, by the larger people. They alone are at home in the steaming darkness of the immeasurable forest into which no other natives dare to enter.

It is a remarkable fact that the Egyptologist Mariette had, before these discoveries, found on an ancient Egyptian monument the portrait of a dwarf inscribed with the word "akka"—the identical name by which they are known at this day in the region where Schweinfurth found them.

Public interest in the pygmy race was rearoused three years ago by the announcement that the party of English naturalists at that time exploring the interior of New Guinea had come across a tribe of these little people in the mountains of that island. The existence of these pygmies in New Guinea was already well known, but fuller accounts of them will be valuable. The Italian traveller Beccari, in 1876, speaks of them as "Karonis," and states that they occupy a chain of mountains parallel to the north coast of the north-west peninsular of the island. D'Albertis, Lawes, and other travellers have seen and described individuals of the pygmy race of the mountains of New Guinea. It is interesting to find that they are described as having the body covered with fine, woolly hair, a feature which is recorded by Schweinfurth, by Stanley, and by an ancient Greek writer, in regard to the Congo pygmies of Africa, and led in former times to the notion that the old traditions and accounts of African pygmies referred, not to human beings, but to chimpanzees!

The Laplanders are the only very small-sized people in Europe, but they run from 5 ft. upwards, whereas the negrites and negrillos run from about 4 ft. to less than 5 ft. The Lapps (of whom there are about 25,000 in Finmark and Lapmark) are a thick-set, round-headed (brachycephalic), dark-yellow race, and have always been credited with powers of witchcraft and magic by their neighbours and by modern sailors. They live in immediate contact with the Finns (both are Mongolian races), who are very tall and have fair hair and blue eyes. Some writers have supposed that the Lapps are the remnants of a small race which was formerly spread over the whole of Europe, and was exterminated or driven out by the larger races. But we have no evidence in favour of this view and strong evidence against it, since we now know the skulls and skeletons of a great number of the prehistoric inhabitants of Europe belonging to the Bronze, to the Neolithic, and to the Palæolithic periods. None of these skeletons belong to an abnormally small-sized race, though the Bronze-age people were smaller than their predecessors and successors. The cave-dwellers of the "reindeer" epoch of the Palæolithic period were big men, with fine, high skulls, and even the earlier Palæolithic men of the glacial period, the man of the Neanderthal, the couple from Spy, and the three recently dug up near Perigueux (of whom I have written in another book),[8] were not diminutive men. It is true they were not tall—only about 5 ft. 4 in. in height—but they were very powerful and muscular, and totally different physically from the Lapps or from any of the tropical pygmy men. It is a remarkable fact that in one cave at Mentone, on the Riviera, explored by the Prince of Monaco, two skeletons have been found belonging to a shortish negro-like race (indicated by the form of the skull), and apparently a little later in date than the Neandermen. We must remember that at that remote date there was continuous land connection between Europe and Africa. There is, in fact, no reason to suppose that a pygmy race ever existed in Europe, though, of course, individuals of exceptionally small stature are often produced, and in some regions the whole population is shorter than it is in others.

A very interesting question in connection with the origin and significance of pygmy races of men is, "Why is any race smaller in size than another?" Every species among the higher animals has its standard size from which only in the rarest cases are there departures. That in itself is a curious fact. How was the standard size determined, and how is it maintained? The whole question lies there. At first sight it seems to many people quite simple to account for "pygmies"; they will tell you that the poor creatures are half-starved and so unable to grow to full size. That explanation does not, however, meet the case, for the African and Asiatic pygmy races are just as well nourished as most of their neighbours. Also if we look a little further we find that the women of every race are smaller than the men, and often much smaller. That is not because they are ill-nourished as compared with the men. And, again, we find very closely similar species of animals existing side by side, one a large species and the other a small one, having the same opportunities of obtaining regular nourishment. There are many instances, but take for example the beautiful Great Koodoo antelope of Africa, with its fine spiral horns, which measures 5 ft. at the shoulder, and the Little Koodoo, a complete miniature of it existing alongside of it, and standing only 3 ft. 5 in. at the shoulder. Take the two common white butterflies of this country, the Large White and the Small White, also the Large Tortoiseshell butterfly and the small. Take the instance of many plant genera of which larger and smaller species are found growing side by side. The difference in size in these cases cannot be traced to any insufficiency of nutrition in the smaller kind.

It is evident that difference of size in animals has some deep-lying cause, which is not merely the greater or less abundance of food. Numerous specimens of a perfectly well-formed elephant, closely allied in structure to the Indian elephant, but only 3 ft. high, are found fossil in Malta and the neighbouring Mediterranean region, and in Liberia a species of hippopotamus, distinct from that of other African regions, is common, which is not bigger than a common pig. Pygmy hogs, pygmy deer, pygmy buffaloes (and many other pygmy animals) are known as thriving wild species, so that it seems clear that there are other causes at work than semi-starvation in the production of pygmy races.

A second suggestion which is sometimes made is that the smaller race, or smaller species of two allied forms, is the original one, and that the larger forms have developed from these and established themselves, without completely destroying the smaller original race. This view has at various times been favoured in regard to the pygmy race of man. There is something plausible in the view that these little men are nearer than normal mankind are to the monkeys, and the fur-like hairiness of their skin has been cited in support of it; but a fatal objection is that the men of the pure pygmy race of Africa and Asia are really not more, but less, monkey-like than many full-sized savages. They have heads and faces nearer in shape to those of Europeans than have the Australians, the Tasmanians, and the negroes. They are more intelligent, shrewd, and skilful than their full-sized neighbours. It is quite possible that they are a very ancient race—more ancient, in their isolation and freedom from complicated customs, habits, and mode of life than other savages—but they are not primitive in the sense of being ape-like in structure or in want of mental capacity.

A third possibility in regard to the pygmy people is that they have been "selected" by natural conditions which favoured the survival of small individuals, and thus established a small race—just as man has established small races of horses, dogs, cattle, or what not, by continually selecting small individuals for breeding, until he has produced such races as the

Shetland pony, the toy terrier, and the Kerry cow. It is necessary to discover or to suggest (if this explanation is to be accepted) what precisely is the advantage, in a state of nature, to a small-sized race in being of small size. The guess is made that the small people can more easily hide, whether in forest or among the rocks and caves of mountainous regions, from aggressive larger-sized mankind. The objection to this view is that though it may explain the present habits and dwelling-places of some of the pygmy race, it is not capable of explaining their first segregation and formation as a distinct race. Another general advantage which small animals have over larger ones of the same species is that if the food of the species is widely distributed but limited in amount, a hundred individuals weighing 5 st. each will secure more of it than fifty individuals weighing 10 st. each. The total weight of individuals is the same, but the smaller series will cover twice the area and have twice as much opportunity to secure the limited amount of food, whilst, in proportion to their size, requiring less. It cannot be doubted that, other things being equal, this obvious relation must tend to limit the increase in size of animals which have to search for their special food, and must favour small races.

Some writers have supposed that small limited areas, such as small islands, favour the production of small races by some mysterious law of appropriateness similar to that which lays down that "who drives fat oxen should himself be fat." The pygmy buffalo of the island of Celebes, the Anoa, is cited as an instance, and the pygmy men of the Andaman Islands as another. But there are plenty of facts which would lead to an exactly opposite conclusion. Gigantic tortoises are found in the Galapagos Islands and in the minute islands of the Indian Ocean, and never on the big continents. Gigantic birds bigger than ostriches abounded in the islands of New Zealand and Madagascar. Some of the tallest races of men are found in the Pacific islands, whilst the tallest European population is that of the north of the island called Great Britain. Probably the real relation of islands to the matter is that owing to their isolation and freedom from the general competition of the vast variety of living things in continental areas, they offer unoccupied territory in which either exceptionally small or exceptionally big races may flourish—if once they reach the island shelter, or are by variation produced there—without competitive interference.

An important consideration in regard to the formation and segregation of a human variety or race is that mankind shows a tendency to segregate in groups, like with like. To a large extent this is true also of animals, but in man it acquires a special dominance, owing to the greater activity in him of psychical or mental influences in all his proceedings. The "cagots" of mid-France are the descendants of former leper families. They remain separated from the rest of the population, and do not now know why, nor do their hostile neighbours. Such "outcast" or "accursed" tribes and family groups

are found also in Great Britain, and throughout the world. Possibly the "pygmies" owe their preservation to this tendency. Virchow regarded the Lapps as a race produced by disease—a pathological product. It is possible that former liability to disease and present immunity from it is the final explanation of the tropical pygmy race. In the United States black pigs are able to eat, without harm, a common marsh herb, the "Red-root" Lachnanthes tinctoria, which kills other pigs. Hence a black race is established, not because it is black, but because, in it, blackness is "the outward and visible sign of an inward and chemical grace"—that is to say, of a physiological or chemical power of resistance to, and immunity from, the poison of an otherwise nutritious plant. Such "correlations" were described by Darwin, and are of extreme importance and interest—far more so than is, at present, recognised by naturalists. I am inclined to the supposition that the obvious outward signs, the round head, bombous forehead, furry skin, and diminutive size of the pygmies are the outcome of an inward physiological condition peculiar to them, which has enabled them to resist disease or to eat certain kinds of food, or possibly to develop great mental acuteness, and so has led to the establishment of these peculiar small people as a race, without their smallness itself having anything to do with their selection and preservation. In that case smallness would be a "by-product," a "correlated" character, not the "effective life-saving" character.
[8] "Science from an Easy Chair," Methuen, 1909.

PREHISTORIC PETTICOATS

After the last great extension of glaciers in Europe, during which nearly all of Great Britain and the North of France and Germany were buried with Scandinavia under one great ice-sheet—and when this ice-sheet had receded, and the climate was like that of the Russian "steppes," cold and dry—there were men inhabiting the caverns on both sides of the Pyrenees. The tract of land which we call "Great Britain" was a part of the Continent of Europe. There was no "English Channel." The Thames and the Rhine opened by a common mouth into the North Sea. The mammoth and the hairy rhinoceros still lingered on in France and the more central regions of Europe. Wild horses, the great ox (Aurochs), the bison, ibex, chamois, were abundant, and the thick-nosed Saiga antelope, now confined to the Russian and Asiatic steppes, was present. The most abundant and important animal immediately north of the Pyrenees was the reindeer. The cave-men of France and Central Europe were a fine race—living by the chase, and fabricating flint knives and scrapers, fine bone spearheads and harpoons, as well as occupying themselves in carving ivory and reindeer antlers, so as to produce highly artistic representations of the animals around them.

They rarely attempted the human face or figure, and when they did were not so successful as in their animal work. They also painted on the walls of some of their caverns, with red and yellow ochre, carbon, and white chalk representations—usually about one-third the size of nature—of some of the most important animals of the chase. They must have used lamps, fed with animal fat, to illuminate the walls, both when they were at work on the pictures and also afterwards, when they exhibited the finished pictures to the less gifted members of the tribe, as wonderful, even magical appearances. It is uncertain to what extent races of men succeeded one another or were cotemporaries in this period in Europe, but there is good reason for attributing the cave pictures to an early occupation of the caves by men who also carved, in ivory and stone, small figures of women resembling the Hottentot Venus—whilst the later occupants made no such statuettes, but carved in relief on bone or engraved it.

This was probably not less than 50,000 years ago, and may well have been much more. Earlier than the date of these Reindeer men (the Magdalenians, Solutrians and the Aurignacians[9]), in the preceding cold, humid period of the glacial extension (probably from 80,000 to 150,000 years ago) these and other caves were occupied by an inferior race—the Neandermen. They could not carve beasts on ivory nor paint, but could make very good and well "dressed" flint weapons, and could make large fires in and about the caves, both to cook their meat and to keep off the wild beasts (lions, bears, and hyenas), who contended with the strange, low-browed Neandermen for the use of the caves as habitations.

On this side of the Pyrenees the Reindeer men have left some wall-pictures, and new discoveries of great importance in the form of rock carvings of human figures as well as pictures and huge figures of horses, etc., are being made in France as I write these lines. But the best preserved and most numerous wall pictures are those of the cave of Altamira near Santander. These comprise some partially preserved representations in yellow, red, white, and black of the great bison, the wild boar, the horse, and other animals. A group representing some twenty-five or more animals (each about one third the size of nature), irregularly arranged, exists on a part of the roof, and others are found in other parts of the cavern. Among the wall-pictures made by ancient cave-men are numerous drawings of human beings in masks representing animals' heads—probably indicating the "dressing-up" in animal masks of priests or medicine men in the way in which we know to-day is the custom among many savage tribes. Twenty-seven of these "decorated" caverns were known in 1910—eleven in Spain, one in Italy, and fifteen in South and Central France—and others are continually being discovered. The most careful and critical examination by scientific men leaves no doubt as to the vast antiquity of these paintings, and as to their dating from a time when the animals painted (including in some cases mammoth and rhinoceros, as well as bison, reindeer, wild boar, ibex, red deer, bear, and felines) were existing in the locality. The covering up of some of the drawings (which are partly engraved and partly painted) by earthy deposits and by encrustations of lime, and the presence in the cave deposits of the worked flints and bones characteristic of the Reindeer men, leave no doubt that these pictures are of that immense antiquity which we express by the words "Quaternary period," "Upper Pleistocene" or "Reindeer epoch."

It is, of course, only in accordance with what one would expect that these pictures are of very varying degrees of artistic merit. But some (a considerable number) are quite remarkable for their true artistic quality. In this respect they differ from the rock paintings of modern savage races—the Bushmen of South Africa, the Australians, and the Californian Indians—with which, however, it is instructive to compare them. Many of them agree in their essential artistic character with the carving and engraving of animals on bone and ivory so abundantly produced by the later Reindeer men. It is also the fact that these Franco-Spanish wall paintings were executed at different periods in the Reindeer epoch. Some are more primitive than others; some are very badly preserved, mere scratched outlines with all the paint washed away by the moisture of ages; but others are bright and sharp in their colouring to a degree which is surprising when their age and long exposure are considered. The French prehistorians, M.M. Cartailac and the Abbé Breuil, have produced a sumptuous volume containing an account, with large coloured plates, of the best preserved of the Altamira paintings—a copy of which I owe to the kindness of H.S.H. the Prince of Monaco, who has ordered the publication of the work at his own charges. This has been followed by an equally fine work under the same auspices, illustrating the wall-pictures of the Cavern of the Font-de-Gaume in the Dordogne, for which we have to thank the Abbé Breuil. A further volume on Spanish Caves has also appeared from the same source in the present year. It is not surprising that the country folk, who, in some of the Spanish localities, have known the existence of these paintings from time immemorial, should regard them as the work of the ancient Moors, all ancient work in Spain being popularly attributed to the Moors, as a sort of starting-point in history. It is, however, very remarkable that little damage appears to have been done by the population to the paintings, even when they exist in shallow caves or on overhanging rocks. No doubt weathering, and the oozing of moisture, and the flaking caused by it, has destroyed most of the Pleistocene paintings which once existed, and it is an ascertained fact that some—for instance, those of Altamira—are breaking to pieces owing to the opening-up and frequentation of the caverns.

It has been remarked that, although these paintings belong to what is called the "reindeer epoch," yet in the cave of Altamira there are no representations of reindeer, but chiefly of bison and wild boar. It is also remarkable that in the case of the painted rock shelters of Calapata (Lower Aragon) and of Cogul (near Lerida, in Catalonia), no reindeer are represented; but on the former there are very admirable drawings of the red deer, and on the latter silhouettes of the bull, of the red deer, and the ibex. In fact, no representations of reindeer have been observed on cave walls or rock-shelters south of the Pyrenees. It is possible that this may be due to the date of the Spanish paintings being a good deal later than that of those French cave-paintings which show reindeer, mammoth, and rhinoceros.

And we have to bear in mind that in the North of Africa (Oran) engraved drawings on exposed rocks are known, which are for good reasons attributed to the Neolithic period; that is to say, they are later than the Reindeer epoch of the Palæolithic period, whilst some are even much later.

In any case we have to remember that there are two very different and possible explanations of the presence or absence either of certain animals' bones or of representations of certain animals in one "decorated" cave and not in another. The one explanation is that animals have succeeded one another in time in Western Europe—changing as the climatic conditions have changed—and that when, in two cave-decorations or cave-deposits compared, the animals are different, the cause may be that the one deposit or cave-decoration is more recent than the other. The other explanation is that (as we well know) at one and the same moment very different animals occupy tracts of land which are only a hundred miles or so apart, but differ in climate and general conditions. At this moment there are wild bears and also wolves in France, but none in England; the elk occurs in Sweden and Russia, but not in the West of Europe; the porcupine in Italy and in Spain, but not in France. As late as the historic period the African elephant flourished on the African shore of the Mediterranean, but not in Spain; now it is not found north of the Sahara at all. So we have various possibilities to consider in comparing the animal pictures on the cave walls of Spain with those found in France, and may well suspend judgment till we have knowledge of a greatly extended area.

I am anxious to draw attention in this chapter to the painted group of ten human figures lately discovered on a rock shelter at Cogul, near Lerida, in Catalonia, and figured and described in the admirable French journal called "L'Anthropologie." These figures are those of young women dressed in short skirts and curious sleeves, the hair done up in a conical mass rising from the sides to the top of the head. Each figure is about ten inches high. The great interest about these drawings is that they are probably tens of thousands of years old, and present to us the women of the reindeer or late Pleistocene epoch. No other such painting of the women of this period is known, and the astonishing thing is that, though these are by no means fine specimens of prehistoric art, yet there is a definitely modern look about the figures and a freedom of touch about the drawing which makes one think at first that the picture is some modern, hasty but clever sketch in silhouette of a number of short skirted school girls at play. The waist is extremely small and elongated, the skirt, or petticoat, bell shaped, and the whole figure "sinuous." One of the figures appears to have a cloak or jacket, but the breasts and legs are bare.

Some three years ago Sir Arthur Evans discovered in the palace of the ancient Kings of Crete coloured frescoes some 3,500 years old representing in great detail elegant young women with greatly compressed waists, strongly-pronounced bustles, and elaborately ornamented skirts. These Cretan paintings of prehistoric young women, both in costume and pose, are like nothing so much as the portraits of distinguished ladies of the fashionable world of Paris exhibited by the painter, Boldini, in the "Salon." It is remarkable that explorers should have found contemporary paintings of young ladies who lived nearly as long before Cleopatra as she lived before us. And it is still more remarkable that those young ladies were "got up" in the same style, and apparently aimed at much the same effects of line and movement, as those which have become the latest fashion in Paris, and may be described as sinuous and serpentine. Not only is that the case, but it is evident that the painter of Knossos, the Minotaur city, and M. Boldini have experienced the same artistic impression, and have presented in their pictures the same significance of pose and the same form, from the tip of the nose to the ends of the fingers and the points of the toes—thus revealing a sympathy reaching across many ages. It seems to me that the same artistic impression is to be detected in the still earlier paintings of the wasp-waisted little ladies of the Cogul rock-shelter in Catalonia. We find here the same sinuous figure with exaggeratedly compressed waist, prominent bosom, and emphasised haunches. But it is many, perhaps forty, thousands years earlier! One is led to wonder whether this type of human female—to-day expressed with such masterly skill by Boldini—may not be at the back of the mind of a portion of the human race—that which populated what are now the shores of the Mediterranean, and probably came there travelling northwards from the centre of Africa. Possibly they brought with them that tendency to, and admiration for, megalopygy which is evidenced by the makers of the earliest known palæolithic cave sculptures (the Aurignacians), and has persisted in some degree ever since in Europe—a tendency and a taste which are on the one hand totally absent in the East and Far East (Japan), and on the other hand have a strong development in the modern Bushmen (and the related Hottentots), an African race, and like the Spanish cave-men, rock painters.

I am able to reproduce here, through the kindness of Sir Arthur Evans and Dr. Hogarth, the keeper of the Ashmolean Museum at Oxford, two very interesting drawings—showing certain features in the dress of women in the prehistoric race which inhabited the island of Crete for some three thousand years previous to the date of these representations, which is about 1600 b.c. They are interesting to compare both with the much more ancient figures from the Spanish cave and with modern female costume. The first is a figure in coloured pottery (faïence), representing either a votary or priestess of a goddess to whom snakes were sacred. The petticoat of this lady is very modern, being long, decorated with flounces (a series of five) and bell-shaped. The dress is further remarkable for a tight ring-like girdle which greatly compresses the waist and emphasises the broad hips. The little statue is about ten inches high, and was found by Sir Arthur Evans at Knossos, the ancient buried city the capital of Crete, in the Later Palace. Its date is that of the close of the Minoan period, namely 1600 b.c. The two figures in Plate IX are copied from frescoes representing acrobatic women from the bull-ring, also from the Later Palace at Knossos, and are a couple of centuries later in date. Religious ceremonies in connection with the worship of the bull (whence the fable of the minotaur) were practised in Knossos, and possibly there was a kind of baiting of bulls and jumping over and away from the infuriated animals such as may be seen at this day in the South of France and in Portugal. Possibly the employment of girls in this sport gave rise to the story of the maiden tribute from Athens to be sacrificed to the Cretan minotaur. The drawings are remarkable for the pose—that of the left-hand resembling an attitude assumed in boxing, whilst the dress—a kind of maillot or "tights"—is gripped round the waist by a firm ring (like a table-napkin ring), the compression of which is no doubt exaggerated. This fresco and many others of extraordinary interest, as well as much beautiful pottery and the whole of the plan of the city, its public buildings, granaries, library and sewers at several successive ages (the remains lying in layers one over the other), were discovered and described by Sir Arthur Evans, who is still at work on the wonderful history and art of these prehistoric Cretans, from whom the Mycenæans of the mainland of Greece were an offshoot.

The point to which I chiefly desire to call attention is that this Cretan people practised compression of the waist, and so have a certain point of agreement with the prehistoric race of Lerida represented in Figs. 24 and 25 and with Boldini's modern ladies. We know from carvings and pottery that the men as well as the women of the Mycenæan people wore a tightly-compressing girdle. The form of figure thus produced—viz. relatively small, flexible waist, and large hips with protruding buttocks—seems to be a less pronounced variety of that of the small ivory figures of Aurignacian age (late Palæolithic) found in cave deposits of France and of that of the Bushmen women. It seems as though the "ideal" female figure or that admired and pictured by these races and by the modern Latin races is the same in its main features, and differs altogether from that admired in the Far East. Such deeply seated tastes may possibly (indeed, not improbably) be due to a common origin of the Mediterranean and African peoples distinct from that of the Mongoloid Asiatic races.

[9] A brief account of the skulls and implements of primitive man, with illustrations, is given in the first series of "Science from an Easy Chair," published in 1910 by Methuen and Co.

MORE SCIENCE FROM AN EASY CHAIR

NEW YEAR'S DAY AND THE CALENDAR

I came across a discussion the other day as to whether it is right to tell children and to let them believe that Santa Claus puts Christmas presents in their stockings, and that Peter Pan really comes in at the window and teaches nice little boys and girls to float through the air. I was surprised that anyone should be so singularly ignorant of child-nature as to hold that children really believe these things. Children have a wonderful and special faculty of "make-believe," which is not the same as "belief." All the time when a child is indulging in "make-believe" (a sort of willing self-illusion or waking dream) its real, though tender, reasoning-power is merely "suspended," and is not offended or outraged. That power can on emergency be brought to the front, and the little one will say, "Of course, they're not real," or "I always knew he didn't really come down the chimney." So that I do not think anyone need be anxious as to doing harm or laying the foundations of future distrust by telling fairy-tales to the very young. If told in the right form and spirit they are received by six-year-old and older children readily and naturally as belonging to that delicious world of "make-believe" which (as one of their own orators, I believe, has said) "children of even the meanest intelligence will not be guilty of confounding with that very inferior every-day world of reality in which we find, much to our regret, that it is necessary to spend so large a part of our time." The power of make-believe is almost limitless, and makes its appearance even in the speechless infant of less than two years old, who will gather fruit from a coloured picture, generously offer you a bite, and pretend to swallow the rest itself. Make-believe must have been a very big factor in the life of the ape-like predecessors of prehistoric man.

Deception in the world of reality is very different from make-believe, and a terrible thing. To the child—deception in regard to real things, whatever excuses adults may put forward in its defence, is well-nigh unforgivable. To be one who never says "it is" when it is not, nor "it will be" when it will not be—that is to be a friend on whom a child rests in perfect trust and happiness.

What have these thoughts to do with the New Year? Merely this, that it is not only with and for children that we make-believe at this season—we all of us, more or less, indulge in a make-believe about the New Year. As the clock strikes its twelve notes at midnight on December 31st, and all the bells of a great city are heard hovering in the air, sending forth their sweet sounds from far and near into the fateful night, there are few of us who have not a feeling that a great event has occurred. A physical change has set in—the Old Year is dead and gone, and the New Year, something tangible, which you can let in at the door or the window—has just come into being, and is there waiting for us. We are, of course, indulging in "make-believe," for there is no New Year, with any natural, noteworthy thing to mark its commencement, starting at midnight on December 31st. New Years begin every day and hour, and it is by no means agreed upon by all nations of the earth to pretend that the 1st of January is the critical day which we must regard as that portentous epoch, the beginning of the New Year. This choice of a day was made by the Romans, and that wonderful man Julius Cæsar had a great deal to do with it; modern Europe adopted his arrangement of the year or calendar. But the Jews have their own calendar and their own New Year's Day, which varies from year to year, from our September 5th to our October 7th. It is, however, to them always the first day of the month Tishri, and the first day of their new year. The Mahomedans took the date of the flight of Mohammed from Mecca to Medina—the night of July 15th, 622 a.d.—as the commencement of their "era," and its anniversary is the first day of their month Muharram and the first day of their year—their New Year's Day. As, although they reckon twelve months to the year, their months are true lunar months, and are not corrected as are those in use by us (as I will explain below); their year consists of 354 days 8 hours, and so does not run parallel to our year at all. Their New Year's day, which began by being our July 16th, was in the next year coincident with our July 6th, then in three successive years it occurred on different days of June, and so on through May, April, and the preceding months, so that in thirty-two and a half of our years their New Year's Day has run through all our months and comes back again to July.

So much for New Year's Days; they are arbitrary selections, and though the Roman New Year's Day, or January 1st, has been precisely defined and fixed by the determination by astronomers of the position of the earth on that day in its revolution around the sun, yet the original selection of January 1st for the beginning of the year seems to have been merely the result of previous errors and negligence in attempting to fix the winter solstice (which now comes out as December 22nd). This is the day when the sun is lowest and the day shortest; after it has passed the sun appears gradually to acquire a new power, and increases the duration of his stay above the horizon until the longest day is reached—the summer solstice (June 21st). Julius Cæsar took January 1st for New Year's Day as being the first day of a month nearest to the winter solstice. The ancient Greeks regarded the beginning of September as "New Year."

Were mankind content with the measure of time by the completion of a cycle of revolution of the earth around the sun—that is the year—and by the revolution of the earth on its own axis—that is the day or day-night ([Greek: nychthêmeron]) of the Greeks—the notation of time and of seasons would be comparatively simple. No one seems to know why or when the day was first divided into twenty-four hours, nor why sixty minutes were taken in the hour and sixty seconds in the minute. The ancient astronomers of Egypt and China, and their beliefs in mystical numbers, have to do with the first choosing of these intervals in unrecorded ages of antiquity (as much as 2000 or 3000 b.c.). The seven days of the week correspond to the five planets known to the ancients, with the addition of the sun and the moon. But the Greeks made three weeks of ten days each in a month. The true year—the exact period of a complete revolution of the earth around the sun—is 365 days 5 hours 18 minutes and 46 seconds. It was measured with a fair amount of accuracy by very ancient races of men, who fixed the position of the rising sun at the longest day by erecting big stones, one close at hand and one at a distance, so as to give a line pointing exactly to the rising spot of the sun on the horizon, as at Stonehenge. They recorded the number of days which elapsed before the longest day again appeared, and they marked also the division of that period by the two events of equally long sunlight and darkness—the spring and the autumn "equinox." It is obvious that if they took 365 days roughly as the period of revolution they would (owing to the odd hours and minutes left out) get about a day wrong in four years, and it was the business of the priests—even in ancient Rome the pontiffs were charged with this duty—to make the correction add the missing day, and proclaim the chief days of the year—the shortest day, the longest day, and the equinox-days of equal halves of sunshine and darkness. In ancient China, if the State astronomer made a wrong calculation in predicting an eclipse he was decapitated.

It is easy to understand how it became desirable to recognise more convenient divisions of the year than the four quarters marked by the solstices and the equinoxes. Various astronomical events were studied, and their regular recurrence ascertained, and they were used for this purpose. But the most obvious natural timekeeper to make use of, besides the sun, was the moon. The moon completes its cycle of change on the average in 29-1/2 days. It was used by every man to mark the passage of the year, and its periods from new moon to new moon were called, as in our language, "months" or "moons," and divided into quarters. It is, however, an awkward fact that twelve lunar months give 354 days, so that there are eleven days left over when the solar year is divided into lunar months. The attempt to invent and cause the adoption of a system which shall regularly mark out the year into the popular and universally recognised "moons," and yet shall not make the year itself, so built up, of a length which does not agree with the true year recorded by the return of the rising sun to exactly the same spot on the horizon after 365 days and a few hours, has been throughout all the history of civilised man, and even among prehistoric peoples, a matter of difficulty. It has led to the most varied and ingenious systems, entrusted to the most learned priests and state officers, and mostly so complicated as to break down in the working, until we come to the great clear-headed man Julius Cæsar.

In the very earliest times of the city of Rome the solar year, or complete cycle of the seasons, was divided into ten lunar months covering 304 days, and it is not known how the remaining days necessary to complete the solar revolution were dealt with, or disposed of. The year was considered to commence with March, probably with the intention of getting New Year's Day near to the spring equinox. The Celtic people and the Druids, with their mistletoe rites, kept New Year also at that time. The ten Roman months were named Martius, Aprilus, Maius, Junius, Quintillis, Sextilis, September, October, November, December. In the reign of the King Numa two months were added to the year—namely, Januarius at the beginning and Februarius at the end. In 452 b.c. February was removed from the end and given second place. The Romans thus arranged twelve months into the year, as the ancient Egyptians and the Greeks had long before done. The months were made by law to consist alternately of twenty-nine and of thirty days (thus keeping near to the average length of a true lunar cycle), and an odd day was thrown in for luck, making the year to consist of 355 days. This, of course, differs from the solar year by ten days and a bit. To make the solar year and the civil or calendar year coincide as nearly as might be, Numa ordered that a special or "intercalary" month should be inserted every second year between February 23rd and 24th. It was called "Mercedonius," and consisted of twenty-two and of twenty-three days alternately, so that four years contained 1465 days, giving a mean of 366-1/4 days to each year. But this gave nearly a day too much in each year of the calendar (as the legal or civil year is called) as compared with the true solar year, agreement with which was the object in view. So another law was made to reduce the excess of days in every twenty-four years. Obviously the superintendence of these variations, and the public declaration of the calendar for each year, was a very serious and important task, affecting all kinds of legal contracts. The pontiffs to whom the duty was assigned abused their power for political ends, and so little care had they taken to regulate the civil year and keep it in coincidence with the solar year that in the time of Julius Cæsar the civil equinox differed from the astronomical by three months, the real spring equinox occurring, not at the end of what was called March by the calendar, but in June!

Julius Cæsar took the matter in hand and put things into better order. He abolished all attempt to record by the calendar a lunar year of twelve lunar months; he fixed the length of the civil year to agree as near as might be with that of the solar year, and arbitrarily altered the months; in fact, abandoned the "lunar month" and instituted the "calendar month." Thus he decreed that the ordinary year should be 365 days, but that every fourth year (which, for some perverse reason, we call "leap" year) should have an extra day. He ordered that the alternate months, from January to November inclusive, should have thirty-one days and the others thirty days, excepting February, which was to have in common years twenty-nine, but in every fourth year (our leap year) thirty. This perfectly reasonable, though arbitrary, definition of the months was accompanied by the alteration of the name of the month Quintilis to Julius, in honour of the great man. Later Augustus had the name of the month Sextilis altered to Augustus for his own glorification, and in order to gratify his vanity a law was passed taking away a day from February and putting it on to August, so that August might have thirty-one days as well as July, and not the inferior total of thirty previously assigned to it! At the same time, so that three months of thirty-one days might not come together, September and November were reduced to thirty days, and thirty-one given to October and December. In order to get everything into order and start fair Julius Cæsar restored the spring equinox to March 25th (Numa's date for it, but really four days late). For this purpose he ordered two extraordinary months, as well as Numa's intercalary month Mercedonius, to be inserted in the year 47 b.c., giving that year in all 445 days. It was called "the last year of confusion." January 1st, forty-six years before the birth of Christ and the 708th since the foundation of the city, was the first day of "the first Julian year."

Although Julius Cæsar's correction and his provisions for keeping the "civil" year coincident with the astronomical year were admirable, yet they were not perfect. His astronomer, by name Sosigenes, did his best, but assumed the astronomical year to be 11 min. 14 sec. longer than it really is. In 400 years this amounts to an error of three days. The increasing disagreement of the "civil" and the "real" equinox was noticed by learned men in successive centuries. At last, in a.d. 1582, it was found that the real astronomical equinox, which was supposed to occur on March 25th, when Julius Cæsar introduced his calendar (not on March 21st, as was later discovered to be the fact), had retrograded towards the beginning of the civil year, so that it coincided with March 11th of the calendar. In order to restore the equinox to its proper place (March 21st), Pope Gregory XIII directed ten days to be suppressed in the calendar—of that year—and to prevent things going wrong again it was enacted that leap-year day shall not be reckoned in those centenary years which are not multiples of 400. Thus Pope Gregory got rid of three days out of the Julian calendar, or civil year, in every 400 years, since 1600 was retained as a leap-year, but 1700, 1800 and 1900, though according to the former law leap-years, were made common years, whilst 2000 will be a leap-year. In order to correct a further minute error, namely, the fact that the calendar year as now amended is 26 sec. longer than the true solar year, it is proposed that the year 4000 and all its multiples shall be common years, and not leap years. This is a matter which, though practical, is of distinctly remote importance. Some people like to look well ahead.

The alteration in the calendar made by Pope Gregory was successfully opposed for a long time in Great Britain by popular prejudice. It was called "new style," and was at last accepted, as in other European countries, but has never been adopted in Russia, which retains the "old style." An Act of Parliament was passed in 1751 ordering that the day following September 2nd, 1752, should be accounted the fourteenth of that month. Many people thought that they had been cheated out of eleven days of life, and there were serious riots! The change had been already made in Scotland in the year 1600 without much outcry. The Scotch were either too "canny" or too dull to "fash" themselves about it.

Let us now revert for a moment to the proceedings of Oriental potentates in regard to astronomers, a class of scientific functionaries whom they have from remote ages been in the habit of employing. It appears that in China there is no attempt to make the civil year or year of the calendar coincide with the astronomical year. The astronomical year is reckoned as beginning when the sun enters Capricorn, our winter solstice, and is thus more reasonably defined than is the commencement of our New Year, which is nine days late. Twelve months are recognised; the first is called Tzu, the second Chou, and the third Yin, and the rest respectively Mao, Chen, Su, Wu, Wei, Shen, Yu, Hsu, Hai. But the calendar year, on the other hand, begins just when the Emperor chooses to say it shall. He is like the captain of a ship, who says of the hour, "Make it so," and it is so. With great ceremony he issues a calendar ten months in advance, fixing as he pleases all the important festive and lucky days of the year. Various emperors have made New Year's Day in the fourth, third, second, first, or twelfth month. It has now been fixed for many centuries in the second astronomical month. I have mentioned above that the ancient Greeks reckoned the New Year as beginning about the end of September. But the reckoning differed in the different States, and so did the names of the months. Although the Greek astronomers determined the real solar year with remarkable accuracy, and proposed very clever modes of correcting the calendar so as to use the lunar months in reckoning, there was no general system adopted, no agreement among the "home-ruling" States.

I have stated above that the official Chinese astronomers sometimes get their heads cut off for not correctly foretelling an eclipse. Illustrating this there is the following story of a visit paid about forty years ago to the Observatory in Greenwich Park by the Shah of Persia of that date. The Persians have many close links with the Chinese, and share their view of astronomy as a sort of State function, in which the Emperor has special authority. The Shah accordingly made a great point of visiting the British State observatory, in company with King Edward, who was then Prince of Wales. Sir George Airy was the Astronomer Royal, and showed the party over the building and gave them peeps through telescopes. "Now show me an eclipse of the sun," said the Shah, speaking in French. Sir George pretended not to hear, and led the way to another instrument. "Dog of an astronomer," said the Shah, "produce me an eclipse!" Sir George politely said he had not got one and could not oblige the King of Kings. "Ho, ho!" said the Shah, turning in great indignation to the Prince of Wales. "You hear! cut his head off!" Sir George's life was, as a matter of fact, spared, but in the course of a year he retired, and was succeeded by another Astronomer Royal. On his appointment that gentleman was astonished at receiving a letter of congratulation from the Shah of Persia. The Shah evidently thought that his bloodthirsty request had been attended to, though with some delay. He proceeded to tell the new Astronomer Royal that he had a few days before writing witnessed a total eclipse of the sun in the observatory at Teheran. This was perfectly correct. The suggestion was that the Teheran astronomers knew their business, and had the good sense to arrange an eclipse when a Royal Visitor wished for one, and so escape decapitation—a course which the kindly Shah evidently wished to indicate to the new and young Astronomer Royal as that which he should pursue in order to avoid the fate of his unhappy and obstinate predecessor. The attitude of the Shah towards science is one which is not altogether unknown in this country.

EASTERTIDE, SHAMROCKS AND SPERMACETI

Most people think of Easter as a Christian festival, but it is really in name and origin a pagan one. The word "Easter" is the modern form of "Eastra," the name of the Anglo-Saxon goddess of spring (in primitive Germanic, "Austro"). The Germans, like ourselves, keep its true pagan name, "Ostern." The Latin nations use for Easter the word Pascha (French, Pâque), the Greek form of the Jewish name for the feast of the Passover, with which it is historically associated by the Christian Church. Terrible quarrels have occurred in early ages over fixing Easter Day and its exact relation to the Jewish calendar. This is the explanation of its being "a movable feast" and of the consequent inconvenience to Parliament, schoolboys, and Bank-holiday-makers at the present day. It must be admitted that when Easter comes as early as it sometimes does those who have but the short spring holiday of the Easter week-end are hardly used. Instead of enjoying the sunny spring weather of Austro, and the flowers and the bursting buds which an Easter at the end of April often gives, they have to put up with the dreary chill of arid March, and this, absurdly enough, is all on account of a mistaken attempt at accuracy made by the Church some sixteen hundred or more years ago in trying to bring the Christian festival into line with the Jewish Passover. If it were desired to celebrate the Feast of the Resurrection each year on the day corresponding astronomically with that indicated in the Gospels, the Astronomer Royal would have no difficulty in exactly fixing the day, making due allowance for the changes of the calendar and for the irregularities of the Jewish year. I do not know what day in what month such a calculation would finally establish as that of the ecclesiastical festival, but the Bank Holiday and the Anglo-Saxon Easter might be dealt with separately, and assigned, once for all, to the end of April, the real "opening," or spring month.

The yellow "tansy cakes" which used to be, and the coloured eggs which still are, given away at Easter throughout Europe, are not of Christian origin, but belong to the Roman celebration (at the same season, viz., April 12th to 15th) of the goddess of Plenty—Ceres. Eggs are the symbols of fecundity and the renewal of life in the spring. They were decorated and given in baskets by rich Romans to their friends and dependents at this season. "Hot-cross buns" are peculiar to England, and no doubt have a Christian significance. They have not survived in Scotland, although Easter eggs are well known there (sometimes they are called "pace-eggs"), nor on the Continent, where "Pascal eggs" are an institution. "Buns" owe their name to the old Norse word "bunga," a convexity or round lump, preserved also in our words "bunion" and "bung." In Norman French it became "bonne," and in the fourteenth century was applied to the round loaf of bread given to a horse; the loaf was called Bayard's bonne (pronounced "bun"). In some parts of England a "bunny" still means a swelling due to a blow.

The April fish, the "poisson d'Avril," is the polite French term for what we call an "April fool." But why a fish is introduced in this connection I am unable to say. The custom of sending people on fool's errands on the First of April is probably due to the change of the calendar in France in 1564; but there is a Hindoo feast on March 31st, when similar jokes are perpetrated. It is called "Huli," which, in accordance with phonetic laws, readily becomes "Fooli." This is probably only a coincidence.

A curious Easter custom in country districts in England used to be (perhaps still is) that called "lifting" or "heaving." On Easter Monday two men will join hands so as to form a seat; their companions then "by right of custom" compel the women they may meet to sit, one after the other, on the improvised throne and be lifted or heaved as high as may be. On Easter Tuesday the women perform the same rite upon the men. Strangers thus assailed have been much disconcerted and have recorded their astonishment in "notes of travel." The custom is said to be a popular degeneration of the celebration of the Resurrection.

An early Easter falls little in advance of St. Patrick's Day, when there is much "wearing of the green" and questioning as to what plant is "the real shamrock." This matter has become so involved and developed by wild enthusiasm, ignorance, and false sentiment that it is difficult to deal with it. A distinguished Irishman once showed me the "shamrock" he was wearing in his buttonhole as "the true" plant of that name. He assured me that he had studied the subject from boyhood and knew well the true and the false. "What is its flower like?" I asked him. "It never has a flower at all," he said. Another injustice to Ireland, one must suppose, or a miracle of St. Patrick's! His "green" was a bit of the small variety of the common clover, Trifolium repens, which, of course, produces the usual tuft of florets or clover-head.

It is true that this plant has now been vulgarly substituted for St. Patrick's shamrock. The shamrock is not really the common clover nor any variety of it. The common Dutch clover and its varieties were introduced into Ireland two hundred years ago from England and are not Irish at all! The true shamrock is the delicate little wood-sorrel, Oxalis acetosella, which has a beautifully formed three-split or trefoil leaf of the most vivid green colour, and a white flower like that of a geranium. It is called "fairy-bell" by the Welsh, and was believed to ring chimes for the elfin folk. It was also greatly esteemed for its acid flavour and for various reputed medicinal and magical properties by the Druids and among the early inhabitants of Great Britain and Ireland. Pliny says it never shelters a snake, and is an antidote to the poison of serpents and scorpions—a good reason for its association with St. Patrick! It had already a reputation and sanctity when, if tradition be true, St. Patrick used its threefold leaf to symbolise the doctrine of the Trinity.

It is much rarer to find the wood-sorrel trefoil with a fourth leaflet than it is to find the clover trefoil so provided. The two plants belong to families widely separated from one another. The ancient architectural decoration of trefoil carving, and also the heraldic shamrock in the arms of the United Kingdom, represent the leaf of the wood-sorrel, and not that of the clover. No doubt there has been some sentimental intention in putting forward the humble, abundant, down-trodden dwarf-clover, the very sod itself of Ireland (really introduced from England) as "the shamrock!" But, as often happens in such cases, truth and the ancient and honourable tradition of a beautiful thing have been wantonly disregarded in order to do business in cheap sentiment. Traders are always ready to take advantage of an ignorant public. Common sprats are called "sardines," the name of another and rarer fish, in order to conceal the fact that they are sprats; clarified horse fat is called "fresh country butter," and Irish regiments are made to decorate themselves with common clover under the delusion that it is the shamrock. Other plants have been from time to time utilised to usurp the title of "shamrock." Thus the small Lucerne clover or medicago is often sold as "shamrock" to Irish patriots, and the watercress has been solemnly pat forward as the true shamrock simply because old writers tell us, as evidence of the barbarous state of the Irish, that they fed upon shamrocks and watercress. The true shamrock (the wood-sorrel) was formerly greatly valued all over Europe as a salad and a flavouring herb on account of its leaves containing oxalic acid. It was used for the manufacture of oxalic acid, which was sold as "salts of lemons" for removing iron-mould. It was the basis of the soup and of the green sauce for fish, in which the dock-sorrel (Rumex) has now taken its place. The name "shamrock" is an old Irish word, written "seamragg," and means a little "trefoil." Curiously enough there appears to be an Oriental word, "shamrakh," which I am told is of

Arabic origin, and also means a trefoil. In English writers from the seventeenth century onwards the Irish shamrock is variously written of as "shamroots," "shamerags" (this and the next following with hostile intent), "shame-rogues," "sham-brogues," and "sham-rug."

I am sorry to say that Shakespeare does not mention the shamrock at all. No Irishman who knows the little oxalis or wood-sorrel could wish for a more beautiful floral emblem of the Emerald Isle, or dream of letting the vulgar Saxon intruder—the dwarf clover—take its place. Perhaps it is the Ulstermen who have set up the foreign "Dutch" clover to replace the true shamrock, the wood-sorrel. These changes are easily made. For instance, "green" is not the original colour of Ireland, but light blue—Cambridge blue!

This chapter is one of varied material, and I now pass abruptly from fresh emerald leaflets to the waxy crystals stewed out of the fat of a monster's head. There has seldom been a controversy so entertaining as that between Dr. Bode (the talented director of the Art Gallery of Berlin) and his opponents, in regard to the age of the wax-bust which he purchased not long ago for £8,000 in Bond Street in the belief that it was the work of Leonardo da Vinci. Science has had its share in the examination of the bust. The last scientific contribution to the matter was the discovery by an analytical chemist, Dr. Pinkus, that the waxy mixture of which the bust is composed consists in definite proportion of spermaceti. Now since spermaceti was not used before the year 1700, the bust cannot (say Dr. Bode's opponents) have been made by Leonardo da Vinci, who died in the early part of the sixteenth century. "Nonsense!" reply Dr. Bode's supporters, "Shakespeare makes Hotspur speak of 'parmaceti,' and it was well known to the doctors of Salerno in 1100 a.d., and probably used by the ancients."

Nevertheless, the opponents of Dr. Bode are right. I am sorry, because Dr. Bode is, in regard to "works of art," a most able expert, and I think it is better that experts should always be right. Spermaceti was known, probably from classical times onwards, as a rare and precious unguent, "resolutive and mollifying," as M. Pomel, "chief druggist to the late French King Louis XIV," says in his treatise on drugs, translated into English in 1737. It was applied as a liniment for hardness of the skin and breasts, and was also taken internally. Shakespeare's reference to it is "parmaceti for an inward bruise." The fact is it was known and used in small quantity before 1700 a.d. in connection with medicine and the toilet, but was not consumed by the thousand tons a year, as it was after the hunting of the sperm whale or cachalot (Physeter mecrocephalus) had been set a-going by the brave fishermen of Nantucket and the Northern Atlantic coast of America in 1690. In 1730 or thereabouts the English and the Dutch also sent out ships to take part in this perilous industry, which is now again, in its dwindled condition, exclusively American. It is the pursuit of by far the biggest and fiercest animal which man has doomed to extinction. Those who enjoy such stories of adventure should read Mr. Bullen's personal narrative, "The Cruise of the Cachalot." It was at the end of the eighteenth century that spermaceti became so abundant in the market that candles of it were manufactured and sold cheaper than those of wax. From about 1860 it was superseded by paraffin and other wax-like products: and it was at its cheapest period, and when it was most widely in use, that Lucas, the English artist, who made many wax busts and statuettes, is known to have mixed it, in the form of "old candles," with beeswax, in order to form the composition which he used in his works. The evidence given by the chemist, Dr. Pinkus, appears to me to be conclusive (even without the evidence of the old clothes stuffed into the hollow of the bust) against the theory that the Bode wax-bust of Flora is more ancient than the nineteenth century, and much in favour of its being the work of Lucas, who is exceptionally known as a wax-modeller of repute sixty years ago, who did use spermaceti.

Spermaceti is a perfectly definite chemical body, which can be recognised without chance of error. It is a combination of palmitic acid and a peculiar hydrocarbon, called (after the whale) "cetyl," and easily forms pure crystals. Before sperm whales were hunted it was obtained in relatively small quantity from individual sperm whales, which by misadventure landed themselves on the coast of France, Spain, or Great Britain, and was eagerly purchased by the apothecaries and perfumers of the great cities of Europe. There are several records of such strange mistakes on the part of the great sperm whale. Only ten or fifteen years ago one was stranded on the Lincolnshire coast, whilst the specimen exhibited in the Natural History Museum was washed ashore at Thurso in Caithness. The spermaceti is found dissolved in the more ordinary oil (or fat), which occupies a huge region above the bones of the upper jaw and gives the sperm whale its barrel-shaped head. It separates on cooling, from the liquid oil, in crystalline flakes, forming great masses, which are purified by re-melting and cooling. In early times the fine waxy, flaky material thus obtained was known in samples of a few ounces, and sold by apothecaries. It was known that it came from a whale, and was believed to be the seed or sperm of that animal, hence its name "spermaceti." M. Pomel, whom I cited above, believed it to come from the brain of the whale called "cachalot." No one would have dreamt in the sixteenth century of mixing this precious stuff with beeswax for modelling purposes. At that date one would as soon have mixed amber with pitch. That reminds me that "grey amber" or "ambergris" is also a product of the sperm whale not to be confounded with spermaceti. It is an unhealthy intestinal concretion like bezoar-stone (see p. 64), only exceptionally produced. It is found floating in the ocean, and is recognised as coming from the cachalot owing to its being largely made up of the horny beaks of cuttle-fish, upon which the cachalot feeds. It is still used in perfumery, and fetches the extraordinary price of four guineas the ounce. A piece weighing 4-1/2 oz. may be seen in Cromwell Road.

Though the oils (or fats) of plants and animals are very similar to one another in appearance, there are a very large number of them differing chemically from one another. Thus the fat or oil of dozens of different nuts and plant-products and of lower animals and fishes, and of sheep, oxen, pigs, dogs, elephants, and men contain different and special chemical substances, corresponding to the "cetyl" which is present in the fat of the sperm whale's head. Many of them have acquired as a result of experience and tradition special value for some special purpose. Several oils have peculiar fitness and great value for oiling delicate machinery; others are used in curing leather, for burning, and for medicinal ointments, whilst a large variety is used as human food.

MUSEUMS

The word "museum" is not one of those which explain themselves and give an indication of what the thing to which they are applied should be, when it has ceased to be what it was intended to be. In ancient Greece the word "mouseion" meant "the place of the Muses"—a grove or a temple—and there was such a place on a part of the Acropolis of Athens, the rocky temple-crowned hill around which the city was built. There were other "museums," or seats of the Muses, in ancient Greece; those on the slopes of Mount Helicon and of Mount Olympus were the most famous. In modern times a picture gallery and art collection, that of the Louvre, in Paris, is called "the Musée," whilst "the Muséum" (the Latin form of the same word) is the name distinctively applied in Paris to the collections of natural history and the laboratories connected with them in the Jardin des Plantes. In London "the British Museum," founded in 1753, originally comprised the national library as well as collections of antiquities and of natural history. In Heidelberg "the Museum" was the name, when I was there, for a delightful club, with a garden. It belonged to the professors, their families, and their friends in the town, and concerts and dances were given in it. It seems that the Heidelberg "Museum" comes nearest to the original meaning of the word as "a seat of Muses," for nearly all those mythical ladies were remarkable for their special patronage of music, dancing, and song.

Who were these goddesses, the Muses, and what were their names? What was the speciality of each, and how do they come to have to do with collections of works of art and specimens of natural history? Two learned "classical" friends whom I lately met in Paris could not help me further than by giving me the names of the first three. I was a little shocked, but the next evening discovered that these goddesses are, in modern times, very generally neglected and ignored. In an extremely amusing play, called "Le

Bois Sacré"—the Sacred Grove (of the Muses)—a name applied jocosely to the Ministry of Fine Arts—I found that the minister of that department was represented as a pompous and fatuous person who completely fails to call to mind, in the course of an eloquent speech, the name of more than one. On ringing for his secretaries and airily asking them to refresh his memory, he did not succeed in extracting from them more than two doubtful additions to his list!

I am able, nevertheless (after due investigation), to put my reader in possession of the facts so unfamiliar to the modern oracles of classical mythology! Briefly, it appears that in the best period of ancient Greece nine Muses were recognised, namely, Calliope, the Muse of epic poetry; Euterpé, of lyric poetry; Erato, of erotic poetry; Melpomené, of tragedy; Thalia, of comedy; Polyhymnia, of sacred hymns; Terpsichoré, of choral song and dance; Clio, of history; and Urania, of astronomy. The last two seem to have very little in common with the addiction to singing and dancing characteristic of the rest, and are the only ones who can be imagined as feeling themselves at home in a modern museum, excepting on those evenings when the authorities use the museum (as is the custom in London) for a "conversazione," enlivened by brass bands and songs.

Apollo was said to be the leader and master of the Muses, but was not related to them. They were in origin the "nymphs" or "genii" of mountain streams worshipped by an ancient bardic race (resembling our own sweet-singing Welsh folk), the Thracians. At first the number of the Muses was indefinite, and they had no names. Then three were named—one of Meditation (Meleté), one of Memory (Mnemé), and one of Song (Aöidé)—a much prettier embodiment of the impression made on a poetical mind by rock-pools and cascades and leafy gorges than the formal and redundant nine of later times. One can associate the primitive three with a museum of natural history; but the later official goddesses, each insisting on her own department of poetry, are too clearly representative of the all-appropriating pretensions of literature in modern seats of learning. They remind me of the enumeration of studies which a dear old head of an Oxford college innocently regarded as complete and reasonable when he assured me that all branches of knowledge were fairly and equally represented on the college staff. "We have," he said, "a lecturer on Greek literature, one on Latin literature, one on Greek history, one on Roman history, one on classical philology, one on modern history, one on mathematics and one on the natural sciences." What more, he asked, could you wish for?

It appears that, without any special reference to the attributes of the Muses, the word "museum" has been adopted in recent times for a building in which collections of works of art and specimens of natural history are housed, and even for the collections themselves—in consequence of the foundation by the Ptolemaic Kings of Egypt of a splendid institution at

Alexandria to which the name museum (mouseion) was given. It included the great library, apparatus for the study of astronomy, anatomy, and other sciences, and collections of all kinds. The most learned men were employed in its management and were lodged there and provided with the means of study and teaching. It was a combination of university, learned academy, and temple, and was the pride of the ancient world. It survived many changes of lordship, but at last the library and collections were deliberately destroyed by Moslem invaders in 640 a.d. The precious manuscripts were served out as fuel for the public baths, and were so numerous that it took some months to consume them! The destruction of the museum of Alexandria marks the commencement of the "Dark Ages"; the ancient culture was dead. Eight centuries of submergence with strange mysterious upfloatings were its fate until the Renascence, when its fragments were recovered, and soon did more harm than good to the fetish-worshipping peoples of Europe.

The first use of the word "museum" in this country for a place in which collections of ancient works of art and specimens of natural history were stored and arranged for exhibition was in the early eighteenth century, when it was applied to the building at Oxford, erected for Mr. Ashmole's collections, presented to the University. This was called "Ashmole's Museum," or the Ashmolean Museum. Previously such a collection and its location were spoken of as "a cabinet of rare and curious objects." "Museum" was occasionally used for what we now call a "study," and even to describe lecture-rooms and library. I have not been able to discover that the word was used in its modern sense at an earlier date on the Continent than in England. The first great typical example of a "museum" was the British Museum, founded in 1753. Montagu House, in Bloomsbury, was purchased by the State to serve as a "repository" (the word used in the Act of Parliament of that date) for the vast collections of natural history made by Sir Hans Sloane, with which were associated certain valuable libraries and collections of manuscripts, of coins, and antique marbles. A large part of the money required for the undertaking was raised by a public lottery, over which the Archbishop of Canterbury, the Lord Chancellor, and the Speaker presided (according to the custom of those days in regard to State lotteries), and it is thus that this remarkable group of great officials became, and have remained ever since, "the Three Principal Trustees of the British Museum." Additional trustees were named (since increased to a total of nearly fifty), and provision was made for the appointment of a principal librarian and other curators of the collections. The Act declared that the collections placed in the "repository" (Montagu House) were to remain there for the benefit and enjoyment of posterity for ever—a provision which until seven years ago was misinterpreted, so as to prevent the sending out of unnamed and unstudied collections of small portable objects like insects, dried plants, and shells, to be named and compared with other specimens, by foreign naturalists. Consequently, there was a great accumulation of specimens unstudied and useless, and a great loss to knowledge. But the late Lord Chancellor (Halsbury) decided that it was not only legally within the power of the trustees temporarily to remove specimens from "the repository" for the purpose of having them named and studied, but actually their duty to do so.

We now very generally recognise in Great Britain, as in other parts of the civilised world, the value and importance of public "museums" in the sense of "repositories of collections of objects of ancient and modern art and of natural history." Museums, as at present existing, may be divided into four kinds, according to the nature of the public or private bodies by which they have been set up and carried on. There are, first of all, national museums maintained and continually increased by the expenditure of a great State, and placed in the capital city; secondly, provincial or local museums, supported by a municipality or by local munificence; thirdly, academic museums, which are those related to the instruction and investigations carried on in a university or a school, and forming part of its regular provision for study; and, fourthly, the museums of private individuals (which as a rule, become eventually transferred by gift or purchase to some existing public museum).

The word "museum" would, and often does, fitly include picture galleries, but very usually in Great Britain a museum is not considered as comprising a picture gallery, and a picture gallery is treated and managed as something distinct from "a museum." The distinction is recognised in London, where we have as separate institutions the British Museum and the National Gallery. Probably the distinct method of exhibiting and caring for pictures, and the very large amount of special knowledge connected with the reasonable employment of public funds in the purchase of these very high-priced objects, as well as the example of private collectors of pictures, are the causes which have led in the past to the complete separation of "picture galleries" from "museums." It is, however, a curious fact that the British Museum (which once possessed some oil paintings, now removed to other public galleries) retains and expends money on its splendid collections of water-colour pictures, drawings, and engravings, whilst in the latter half of the last century (in opposition to the custom of separating pictures from other museum objects) there grew up in London, under the State Department of Education, a vast collection of all kinds of works of art (pottery, furniture, lace, metal-work, etc.) of all countries and ages, including pictures, which is now sumptuously housed in the Victoria and Albert Museum.

Though I propose to write here with special reference to "museums," in the more limited sense as repositories of objects which are the bases of our knowledge of the history of man and his arts, and as the storehouses of specimens which in the same way are the material by the study of which we arrive at a knowledge of the history of the earth, and of the living things which have existed, and of others which still exist on its surface—yet it is obvious that the general purposes of all collections of interesting objects (including even pictures) and their arrangement for public use and benefit must be the same, although there are special purposes in view in regard to some collections which do not exist in regard to others. Not long since Mr. Claude Phillips ably set forth some of the principles which should guide the arrangement and exhibition of objects in an art museum, and criticised the plan at present adopted in the Victoria and Albert Museum. As I hold views in regard to the arrangement of natural history museums which are very similar to his, I think it may be useful to explain here what they are.

I may point out that nearly every branch of knowledge should have—in a civilised well-provided community—its collection of material objects, either specimens, models, or ancient examples and remains, which should be "records" to be religiously preserved for future reference and comparison by expert students, whilst others should be there to serve as demonstrations of "great" facts of nature or of human art—direct and straightforward appeals—to the ordinary intelligent (but not specially learned) man. You might well have (what does not at present exist!) a museum (in the modern sense) of astronomy, containing models of the solar system showing the relative distances and sizes of the heavenly bodies—as well as modern and ancient astronomical instruments, and the records obtained by their use. Again, you might have (and to some extent such museums exist), at the other end of the scale in dignity and age, a museum illustrating the history and present developments of the smelting of iron and other metals, their purification, their alloying, and properties—as also a museum of paper-making and one of the steam engine and its modern rivals. In such cases the purpose of the museum would be plain enough and comparatively easy to carry out.

Most museums which have come into existence within the last 200 years suffer from the fact that they are mere enlargements of the ancient collector's "cabinet of rare and curious things," brought together and arranged without rhyme or reason. No one has ever attempted to say what is precisely the aim and intention as a public enterprise of any of our great museums, and accordingly there has been no consideration, discussion, or agreement as to the methods of collection, selection, arrangement, exhibition, and storage of the objects assembled within their walls. Thousands, even millions of pounds, have been expended on the building of museums, on the purchase of specimens, on cases and cataloguing, and on the salaries of directors, and keepers, and assistants, yet the museums remain, so far as any declaration of purpose and principle is concerned, mere "repositories," as in the words of the old Act of Parliament constituting the British Museum—for the use and enjoyment of the public, it is true, but without any expression of a conception of how that use and enjoyment is to be limited so as to make them something better than a dime-show, or how any serious purpose is to be achieved by their costly housing and up-keep. No doubt various directors and keepers have from time to time shown intelligence and laboured to make museums not only places of enjoyment and "edification," but also the means of increasing knowledge and rendering service to the State. But the scope of our public museums, and the principles and methods by which it may be realised, have never been agreed upon, and consequently are not definitely recognised by the State nor by the curiously ill-chosen committees of managers, or trustees, to whose tender mercies the ultimate control of these institutions is confided—apparently by haphazard or misapprehension.

The notion of a town corporation, or of the central government at this or that date, has been that museums are best controlled and public money expended in connection with them by persons who know nothing about the real importance of the collections, and receive no guidance from any scheme or statutable declaration of specific purpose drawn up by a competent authority. I will endeavour to state what those purposes should be.

When one tries to estimate what is really the value to the community of public "museums," one is led inevitably to the conclusion that their most important purpose—whether they are museums of natural history, of antiquities, or of art—is to serve as safe and permanent "repositories" (the old word used in the British Museum Act of 1753) for specimens which are costly and difficult to obtain—not to be either "picked up" or readily "housed" by everybody, and at the same time of real importance as "records." The first and most commanding duty of those who set up and maintain a public museum is to preserve actual things as records—records of the existence in this or that locality of each kind of plant and animal, records of the former existence of extinct plants and animals, with irrefragable certainty as to the locality and the exact strata in which they were found—records of prehistoric man, his weapons and art, and of the animals found with them, records of modern times. Everyone is familiar with this duty of the State and of local public bodies, when it is a matter of preserving written and printed records. They are preserved in various public offices and libraries, and are continually being studied by experts (volunteers or official) and copied in print, so as to furnish us with accurate knowledge of the past.

It is the first and leading business of museums to collect and preserve, with great accuracy as to the locality and circumstances in which each was found, the actual concrete things which are the records of nature, and of the various stages of man's art and industries in every region of the world, just as a library or the Record Office preserves manuscripts and printed documents and books. Collections of such specimens are often made by private individuals, and become too cumbersome for him or his heirs to keep in order. They are then frequently given to a public museum, and I regret to say in many provincial museums are neglected and become mere rubbish, even if they were not so when first given. Often such gifts are rubbish before they are received, and should never have been accepted. But in a great many instances the local museum of a country town is nothing but a rubbish-heap, because the townspeople will not spend the money necessary to obtain the services of a capable curator and to provide cases, labels, catalogues, and attendance. The town councillors usually know nothing about the museum or the value of the objects gathered there, and do not recognise the duty of making it an orderly and carefully tended storehouse of the records of Nature and antiquity of the neighbourhood. Too frequently the town museum is made the means of gratifying the vanity of some local collector, who hands over all sorts of ill-chosen, badly preserved specimens to its ignorant guardians, and is advertised by labels on the cases and by votes of thanks, whilst valuable records placed there in a previous generation are swept into a corner or broken and cast into the cellar in order to make space for the new rubbish!

Unless funds are found to place a specially educated man at the head of a local museum, the museum had better be shut, and such of its contents, as may be desired, offered to one of the big city museums or to the National Museum in London. It is no child's play, maintaining and guarding efficiently a museum which contains "records." It would be a good thing were a committee of naturalists and antiquaries to visit the local museums of the United Kingdom and report on the efficiency of their guardianship and the state of the treasures which they contain. I know two provincial museums very well in which extremely valuable records of prehistoric man and of wonderful extinct animals—found in the neighbourhood and preserved by those who established the museums fifty years ago—are utterly neglected and destroyed by loss of the labels and mixing up of the specimens, in consequence of the death of the persons originally interested in the museum and of the refusal of the town councils to find money to pay for the care of the collections. There can be little doubt that in the present state of local interest in such matters all really important record specimens should find their way to the British Museum in London, where, if accepted, their preservation, so far as it is humanly possible, is assured. That is the distinctive and most creditable feature of our great State-supported museum. At the same time it seems obvious that the records of a provincial area can be, and should be, kept in the county town museum, with a detail and completeness impossible elsewhere, and that it should be the pride of the county to be able to show to a stranger full records of the distinctive features of its natural history and antiquities.

It is clear that whatever failures in this respect may be inevitable in those hopelessly starved and mismanaged "museums" at present surviving to bear witness to the decay of public spirit and intelligent culture in our country towns, the prime duty of the great London museum is to preserve "records" with the greatest nicety and readiness for reference, whilst the duty of actively adding to these records from all parts of the Empire, and, therefore, of the world, and that of minutely studying and reporting upon the collections so obtained and guarded, follow as a matter of course. These collections are the absolutely necessary foundation for the building-up of our knowledge of Nature and of man. We can never say that this branch of scientific knowledge is valuable and that another is a mere fanciful pursuit. Every year it becomes more and more clear that unexpectedly some apparently insignificant piece of detailed scientific knowledge may become of value to the State and to humanity at large. Everyone knows that geology has a great practical value in mining, water supply, and various kinds of engineering, also that botany, as represented by the great State institution at Kew, is of immense value to those who introduce useful plants from one part of the world for cultivation in another. But of late we have seen that entomology—"bug-hunting" as it is scornfully termed—is a science upon which hang not only the revenue of an Empire, but also the lives of millions of men. Destructive insects must be known with the utmost accuracy in order to stop their injury to crops in the distant lands which they inhabit, and also in order to check the diseases carried by them which sweep off vast herds of costly cattle. The mosquitoes and the tsetze flies have been, only recently, proved to be the causes, the carriers, of diseases— malaria, yellow fever, and sleeping sickness—which annually have killed hundreds of thousands of men, colonists as well as natives. I was able to bring together at the Natural History Museum collections of mosquitoes from every part of the world, amounting to thousands of specimens and to some hundreds of kinds. The study of these and of the tsetze flies by skilled entomologists employed in the museum has been a necessary part of the steps now being taken everywhere to preserve human population from the attacks of certain deadly kinds among them, distinguished from the others which are harmless.

Thus, then, it seems that the first and most important purpose for which great "museums" exist is that of "the making of new knowledge"—the increase of science—by furnishing carefully gathered and preserved "specimens" of all kinds, and by working out the history and significance of those collections. But there is a second and distinct purpose which is often ignorantly put in the first place. It is of less importance and quite unlike the first in the methods necessary for its attainment, and yet is conveniently and satisfactorily carried out in conjunction with the first. This second and distinct purpose is the exhibition of such portions of the collections in a museum as are suitable for exhibition (only a smaller portion are so) in public galleries, so chosen, arranged, lighted and labelled as to afford to the public at large the maximum of enjoyment and edification. This is, as it were, a readily accessible enjoyment given to the public in recognition of the large sums of public money expended on the severer and less easily appreciated enterprise of the museum. The public galleries of a museum, whether of natural history, antiquities or art, should not contain the bulk of the collection, but only special things, carefully selected, and equally carefully placed in case or on wall, with artistic judgment as to space-bordering and colour of background, and with scientific perfection of illumination, so as to produce the "just" impression on the leisurely visitor. The public "exhibit" should be arranged so as to draw attention to a series of important facts of structure or quality clearly shown by the specimens, whether they are natural products or works of art, and these facts should be described in printed labels fully, and the reason for attaching importance to them explained at sufficient length. The man who arranges the public galleries (as distinct from the closed study-rooms) of a public museum, should have a special gift of exposition in plain language, and be able to separate (both in regard to his words and to the specimens he selects) the essential from the non-essential, the significant from the redundant.

It is important to make a complete distinction between an exhibition intended for the general public and that intended for advanced students in schools, colleges and universities. The confusion of these two kinds of exhibition is the cause of the failure of many museums and of the dislike with which most people regard a visit to them. The public museum—metropolitan or local—should not include in its purpose the "academic" instruction of schoolboys and university students. That requires a different kind of museum, which is (or should be) provided by the school or university, though, of course, the students should also visit the more popular museums. The funds and staff and space required for the one are not sufficient for both. If both are attempted, the unpopular academic, or scholars', exhibition will get the upper hand and suppress the other, since it is a far easier thing to carry out successfully (for the class aimed at) than is the carefully planned exhibition intended for the "edification" of the greater public. The university museum aims at imparting a much greater amount of detailed and elaborate information than does the great public museum, and requires from the student who uses it a special previous study of the subject, and an exceptional amount of attention and pains in examining the objects exhibited.

Too many of the public museums of Europe aim at the "instruction" of the special student rather than at the "edification" of the general public, whilst most aim at nothing at all except showing, without explanation or comment, a vast mass of specimens or pictures, at the sight of which the patient but bored public gapes with wonder. The public galleries of the Natural History Museum in London have been arranged more distinctly with a view to the edification of the public than those of any other museum which I know. But they still contain too large a number of specimens, and still require an immense amount of work in weeding, selection and labelling, and in deliberately making the specimens exhibited tell a tale which is worth remembering, and can be remembered. Except in the case of the larger specimens, and especially those of fossilized skeletons and shells of extinct animals, it must be remembered that the bulk of the specimens (and, indeed, all the valuable skins of animals and birds, and the vast series of insects and such small things) in that, as in every other large museum, are contained in cabinets protected from the destructive action of light, and arranged for the most part in rooms to which access is obtained only by serious workers after special application. The fishes and other animals preserved in alcohol are kept in a special fire-proof "spirit-building."

A provincial public museum, even if it does not aim at the guardianship of important local "records" of natural history and antiquity, should aim at the edification of the public—the grown-up public—and not at the instruction of school children. The notion that museums are meant for children, which exists, I am sorry to say, even in regard to so splendid and expensive a display of wonderful things as that to be seen at the Natural History Museum, is due to the bad tradition justified by the condition of other museums, where a child may enjoy being astonished, but a grown-up person can take in nothing which appeals to the intelligence. A new city museum is, it is reported, to be established at Birmingham. We may hope that it will not contain the usual unsatisfactory series of badly stuffed exotic animals, birds, and reptiles, and trophies of South Sea islanders' clubs and spears. It should contain first-rate specimens of the living and extinct fauna of Warwickshire, and specimens of foreign animals carefully selected to compare with them and throw light on them; also local prehistoric and antiquarian specimens, illustrated by comparison with the work of savage and remote races. The excellent suggestion has been made that it should contain specimens of the insect-pests of Warwickshire crops. It should also exhibit the minerals from which manufactories of Birmingham draw their metals, and should show the stages of their preparation. It should appeal, not to the boys and girls of Birmingham in the first place, but to the adults, and to do this it should be placed under the care of a really first-rate and ingenious man, who might possibly do for the Birmingham Museum what skilful arrangement and sound knowledge have done for its Art Gallery—an institution intended to appeal not to school children, but to the reasonable adult population of the city.

The principle of exhibiting permanently in public galleries a portion of our great national collections and of preserving another and larger portion in smaller rooms, where they can be more closely but not less carefully disposed and brought out into perfect light and position when required, should be applied to collections of pottery, metal-work, carving, embroidery and such objects, and also to pictures as well as to collections relating to natural history. The chief reason for this is the enormous space required in order to place permanently "on exhibition" all the objects contained in our national art collections, which are continually growing. The vast size of the galleries required, if the entire collections are to be exhibited so that the public may walk in and see anything and everything in it, permanently displayed on walls or in cases—entails gigantic and ever-increasing expenditure of public funds.

But this is not the only objection to these great galleries. The multitude of objects—it may be of pictures—exhibited creates a state of mind in the visitor which prevents his enjoyment of the works of art so exhibited. He is overwhelmed by the vastness of the series offered for his examination and confused and distressed by the close setting of things which require isolation and appropriate surroundings each in its own special way, if they are to be duly appreciated. Not only this, but pictures, as well as other works of art, are, in consequence of the necessity of placing them all in the great public galleries used for the purpose, rarely placed in the most favourable conditions of lighting, and are very often so ill-lighted as to lose all their beauty even if they are not nearly invisible. More public money would be available for the proper care and study of works of art were less spent on the land, building and up-keep necessary for huge galleries.

The desirability of separating a large unexhibited portion from the well-chosen and well-shown exhibited portion of works of art, exclusive of pictures, is, I believe, generally admitted. In the case of pictures the opinion has been expressed that there would be great difficulty in managing a reserved unexhibited portion of our national collections so that the pictures could be properly cared for and yet readily brought into view when required. One can well believe that a similar difficulty was anticipated when it was first proposed to keep books on shelves instead of on tables. Those who take this objection have overlooked the resources of modern engineering. Reserved pictures could be affixed in perfect security in appropriate groups on large screens, and these disposed, like the scenery above a stage, upright and in series, each screen 4 ft. distant from its neighbours. There could be three or four floors of such closely packed screens arranged in two rows, twenty in a row. On a lower floor there would be provided a room with the most perfect light possible for seeing, enjoying and studying a single one of these screens. They would all be numbered and the pictures on each catalogued. A person duly authorised and approved desires to see such and such a picture. He is given a seat in the special exhibition room. The attendant or assistant in charge touches the appropriate button, and by simple electric-lift machinery the screen upstairs carrying the desired picture travels automatically into position and then gently descends into the special exhibition room. There the other pictures on the screen may be, if it be so desired, covered by drapery, the light may be varied in intensity or direction, and, in fact, the most perfect examination of the picture in question may be made. When another button is touched, the picture-screen returns automatically to its place upstairs.

It seems to me that in the case of the growing collection of pictures known as "The National Portrait Gallery," this treatment would not only avoid the necessity of constantly providing new galleries for new acquisitions—but would enable the Trustees to separate those portraits, which are of more general interest and suitable for permanent exhibition in a good position, from less important portraits, which nevertheless must be acquired and preserved as public records. From time to time special groups of the reserved or unexhibited portraits might be put for six months in one of the public rooms—thus providing a change and variety of interest for the general public.

The same plan might be adopted with regard to the pictures in the National Gallery—though no doubt a large number of splendid pictures would be permanently placed in the exhibition rooms. Three things should be remembered in regard to the disposal of these pictures: Firstly, that not one in a hundred among them was intended by the painter to be hung in a gallery closely side by side with other pictures; secondly, that no picture should be exhibited in a public gallery unless it is worthy of the best lighting and surroundings; thirdly, that it is reasonable that the expert and the student should be asked to take some special trouble in order to see special pictures not on public exhibition, and that "the man in the street" who says that he likes to walk in and see all his pictures at any time and without any trouble, will value his collection more when he can only see some of it on special occasions.

The heavy and sometimes fragile character of the "frames" affixed to large pictures has been made an objection to the proposal that they should be fixed to screens moved by electric gear. I cannot venture to discuss the subject of picture frames here. I am aware that it is a very serious and important subject, and that a great deal of the effect of a picture depends on its being bordered by a frame of sufficient size and dignity and one which is really and artistically fitted to allow the finer qualities of the picture to become apparent. How often is such a frame seen? Who is there who has an adequate understanding of picture-frames as adjuncts to, or necessary accompaniments of, great pictures? The splendid carved and gilded wooden frames of some great pictures have a value of their own as examples of design. But how many of them are really suited to the picture which they surround? How much attention has been given by art experts to the question of the best possible "exhibitional" surroundings—nearer and more distant—for this, that and the other, among the great pictures of Europe?

THE SECRET OF A TERRIBLE DISEASE

This generation, which is so thankless to the great discoverers of the causes of disease, so forgetful of the epoch-making labours of the English sanitary reformers of last century, has not seen nor even heard of the awful thing once known as "gaol-fever." A hundred years ago it was as dangerous to the life of an unhappy prisoner to await his trial in Newgate as to stand between the opposing forces on a battlefield. Gaol-fever attacked not only the prisoners, but the judge and the jury and the strangers in the court. The aromatic herbs with which the hall of justice was strewn were supposed to arrest the spread of the terrible infection, and it is still customary to provide with a bouquet of such plants the judge who presides at a "gaol delivery." The inexorable ministers of justice, who, seated high above the common herd, and clad in their ancient robes of office, were about to deal shameful death to the guilty wretches brought from the prison cells, were often themselves struck down by the Angel of Death moving invisibly through the court. The "black assizes" were not isolated, but repeated occurrences in our great cities. Typhus fever was the name given by the learned to this awful pestilence. There was a mystery and horror surrounding it which paralysed those who came into contact with it, and produced something like consternation. Men fled in terror from the infected buildings, business was arrested, the universities deserted, palaces left empty, and the dying abandoned to their misery when it appeared. There was a feeling that some deadly unseen power was present, irresistible and malignant.

It is only to-day—in fact, within the last two years—that we have learnt what that unseen power was. The Angel of Death which moved through the Old Bailey Sessions House in bygone days was, indeed, a living thing. It passed silently and unseen from the prisoner to the warder, from him to the usher, thence to the bar—the jury and the exalted judge. It had no wings,

169

yet it moved slowly and surely carrying black death with it. This terrible and mysterious assassin has at last been unveiled. The shroud of concealment has been torn away and there the dire monster stands—naked, remorseless and hideous. It is of small size, though it makes us all shrink with horror and disgust. It has six claw-like legs and no wings. It is, in fact, neither more nor less than the clothes louse, the Pediculus vestimenti. The filthy, crowded condition in which the prisoners were kept, and (let us well remember and reflect thereon) the personal want of cleanliness of judge, jury, barristers and ushers, rendered the existence of the little parasite and its effective transference from man to man possible. Those pompous emblems of authority, the horsehair wigs—those musty robes of unctuous dignity—were full of dirt, and harboured the wandering bearer of typhus infection. Gaol-fever was due to dirt; its infecting germs were distributed by loathsome insects.

It is an interesting and really instructive thing to pass in review the gradual process by which the cleanliness of the population of Western Europe has advanced, and to observe that, consciously or unconsciously, the end pursued has been, step by step, the removal from man's body outside (and inside), from his clothing, from the water he drinks, from the food he eats, from the air he breathes, and from the surfaces with which he necessarily comes into contact, of injurious parasites and hurtful living things which lurk in dirt and rubbish. At first the larger and more obvious hurtful creatures—snakes, rats, mice, scorpions, blow-flies—were eliminated by some elementary attempts at removal of rubbish and kitchen middens. Then ticks (which African savages still do not trouble to remove from their bodies) and later fleas and bugs became unpopular; lice were long regarded as inevitable, and even beneficial, and by some populations and by part of the most civilised at the present day, are still, not merely tolerated, but favoured. In a country school in France a child who was found to be afflicted in this way was the daughter of the local medical practitioner. She remarked, "Oh! Ce n'est rien; papa dit que c'est la santé des enfants"! Parasitic worms of various kinds, though they often cause disease and death, are accepted and tolerated even by the most refined and luxurious, who risk infection rather than submit to the precaution of abstention from raw vegetables and fruits, or to the expenditure of trouble in cleansing those nests of infective germs. It is only within the last thirty or forty years that such cleanliness of body and of clothing and of house-fittings as will banish parasitic insects has become at all general. The common house-fly is still tolerated, although it is a notorious carrier of dirt and disease, and is bred by dirt and dirt only, its eggs being hatched in old stable manure. The diminution of late years of house-flies in London houses is simply and solely due to legislation compelling the removal of horse manure from the "mews" so frequent at the back of London streets. Egyptian natives still allow flies to gather on their eyelids without protest.

Of the bacteria and similar microscopic germs of disease—to which all our infective fevers are due—we have only become aware quite recently, within the half-century. Before they were known, cleanliness and the destruction of putrescible matter in man's surroundings had, it is true, been urged by sanitary reformers. Disinfectants and antiseptics were deliberately made use of for this purpose in the mid-Victorian period, when carbolic acid and chlorinated lime were established in the place of those feebler destroyers of the germs of putrefaction and disease—namely, the extracts of aromatic herbs or the essential oils themselves. These, as perfumes and unguents, really served, not merely to gratify the olfactory sense, but to destroy by their chemical action the germs of disease. Men tolerated gnats and their bites (mosquitoes as we prefer to call them in order to delude ourselves into the belief that they are not British) until it was discovered that they, and they only, carry the parasitic germs of two deadly diseases—malaria, or ague, and yellow fever. Now we shall destroy the pools in which they breed, just as we are destroying the manure heaps in which the house-fly breeds. When we look over the list it is really astonishing how much remains to be done, even in England, in establishing increased cleanliness and freeing ourselves from the murderous tyranny of parasites. It is a simple but horrible fact that the poorest class in our big cities still swarms with vermin. And not only are the poor in great cities thus afflicted. The recent compulsory medical inspection of school children has shown that in some of the smiling rural districts of England 80 per cent. of the children have lice in their heads. Everyone should help to gain further cleanliness and freedom from this form of oppression.

In the middle of the nineteenth century, England alone, and with absolute conviction and determination, demonstrated to the civilised world the beneficial results in diminishing the death-rate of large towns, to be obtained by cleanliness, the destruction or removal from man's body and surroundings of organic "dirt," viz. his excreta, the exudations and exuviations of his body, the waste and fragments of his food. The names of Rawlinson, Chadwick and Simon remain as those of the prime movers in that legislation which has given us improved water supply, sewerage, removal of dust heaps, clearance of cesspits, cleansing of houses, and prevention of over-crowding. Yet there are writers who, in ignorance and infected with the modern madness which makes half-educated Englishmen presume to teach where they have yet to learn, and to pose as prophets by belittling and running down, without regard to truth, their own country and its finest efforts in the cause of civilisation, actually declare that Germany has led the way in this matter. This is the very reverse of the truth. Foreign countries are, in this matter, following long in the wake of England. There are no cities in the world so healthy as British cities. Practical measures of cleansing, faithful activity in destroying dirt and preventing over-crowding, enforced by legislation, have reduced the death-rate of our great centres of population in fifty years by more than one third—that is to say, from something like 29 per 1,000 to something like 18 per 1,000. No other country can show such a result.

Gaol-fever, spotted or putrid fever, or typhus fever has practically ceased to be a regularly occurring disease in the West of Europe. The last cases in London were, I well remember, in a poor district near the Marylebone Road about thirty years ago. A very few cases have appeared since, in the over-crowded and poorest districts of our largest cities. Beleaguering armies and beleaguered cities suffered from it as late as in the Crimean War, but we may now fairly say that it has disappeared from our midst. It, however, still abounds in Russia and her eastern provinces, and in Algeria, Tunis, and Morocco. It is a disease of cold and temperate climates rather than of the tropics.

In the last century typhus was distinguished definitely and clearly from "typhoid" or "enteric" fever, and from "relapsing" or "famine" fever, with which it had previously been confounded. The bacterial germs causing enteric and relapsing fevers are now known, and have been isolated and cultivated, and the mode in which they are conveyed into the body of a previously healthy patient is ascertained. But until the past year we knew neither the parasitic germ which causes typhus fever nor the mode by which it passes from one individual to another. A vague idea that it was spread through the air prevailed. Typhus is remarkable for the frequency with which the nurses and doctors attending a case become infected. About 20 per cent. of those attacked by it die, but in persons above forty-five years of age the mortality is much greater—about half succumb.

Dr. Nicole and his colleagues of the Institut Pasteur in Tunis have recently had the opportunity of studying typhus there. They found that the ordinary local monkey could not be made to take the disease. But a drop of blood of a typhus patient injected into a chimpanzee (which is far nearer akin to man) produced the disease after an incubation period of three weeks. This fact was definitely established. From what is now known as to relapsing fever, malaria, yellow fever, plague, and sleeping-sickness, it seemed probable that some migratory insect must be the carrier of the typhus infection from man to man. The typhus patients brought into the hospital at Tunis were carefully washed before admission, and no infection of other patients or nurses took place in the wards, although the cases were not isolated, and bugs were abundant. The only cases of infection which occurred were in persons who had the duty of collecting and disinfecting the clothing of the patients when admitted. This seems to exclude the bug as a carrier. The flea is excluded by the fact that in the phosphate mines of Tunis the flea is abundant, and bites both natives and Europeans. Yet when typhus fever broke out among the miners—although all were equally bitten by the fleas—no European was infected. The indication, therefore, was that if any insect is the carrier, it is neither the flea nor the bug, but probably the clothes-louse. Although the smaller monkeys cannot be directly infected with typhus fever from man, it was found that (as with some other infections) the bonnet monkey was susceptible to the infection after it had passed through the chimpanzee. Experiments were, therefore, made with clothes lice taken from a healthy man, and kept for eight hours without food. They were placed on a bonnet monkey which was in full typhus eruption. A day afterwards they were removed to healthy bonnet monkeys with the result that the healthy bonnet monkeys developed typhus fever. There is thus no doubt whatever that typhus fever can be carried in this way from bonnet monkey to bonnet monkey. The whole history of typhus fever fits in with the carriage of the infection in the same way from man to man, and not with the notion of an aërial dispersion of the infection.

The fact that typhus only exists in very dirty and crowded populations, and that it has disappeared where even a moderate amount of cleanliness as to person and clothing has become general, coincides with the possibility of the body louse as carrier. This little parasite is known to be a wanderer, and is gifted with a very acute sense of smell. An individual placed in the centre of a glass table invariably walked, guided by the scent, towards the observer, at whatever position he placed himself. Sulphurous acid is a violent repellant of these creatures. Not only will it kill them if they are exposed to its fumes, but traces of it drive them away. Hence doctors and nurses who have to handle typhus patients or their clothes have only to wear a small muslin bag of sulphur under their garments, or to rub themselves with a little sulphur ointment in order to be perfectly guarded against infection; the louse will not approach them, nor remain upon them should it accidentally effect a lodgment.

It is not always obvious at once in what way a knowledge of the mode of carriage of a deadly disease can be of service to humanity. But in this case it is strikingly and triumphantly clear. In the vast poverty-stricken population of Russia typhus is still common. Public medical officials attend these cases, and the Russian Government keeps a record of the annual deaths of its medical staff, and of the causes of their deaths. In the first six months of last year 530 Russian medical officers died, and twenty-four of these deaths were caused by typhus fever acquired by these devoted public servants in attendance upon cases of that fever. Henceforth they will make use of sulphur or sulphurous ointment to keep the little infection-carriers at a distance, and not one medical man or nurse will catch the disease, still less be killed by it.

A remarkable fact in this history is that the actual parasitic germ which causes typhus, whether a bacterium (Schizophyte) or a protozoon, has not been detected, although the louse has been shown to be its "carrier." The same is true of yellow-fever: we have not seen with the microscope the microbe which produces it. But we know with certainty that the gnat, Stegomya fasciata, and no other, is the carrier of the unseen germ, and that we can obliterate that fever by obliterating the gnat. So, too, although we know how the infection of rabies acts, and how it is carried, yet no one has yet isolated and recognised the terrible infective particle itself. There is a very high probability that in these cases, and also in cancer (where as yet no specific infective germ or parasitic microbe has been detected), such an infective microbe is nevertheless present, and has hitherto escaped observation with the microscope on account of its excessive minuteness and transparency.

CARRIERS OF DISEASE

It has now been discovered that a great number of human diseases are caused by microscopic parasites, which are spoken of in a general way by the name invented by the great Pasteur, viz. "microbes." Wool-sorter's disease, Eastern relapsing fever, lock-jaw, glanders, leprosy, phthisis, diphtheria, cholera, Oriental plague, typhoid fever, Malta fever, septic poisoning and gangrene have been shown to be caused each by a peculiar species of the excessively minute parasitic vegetables known as bacteria (or Schizophyta). Others, for example, malaria and sleeping sickness, have been shown to be caused by almost equally minute microbes, which are of an animal nature, and similar to the free-living animalcules which we call Protozoa, or "simplest animals," whilst a third lot of diseases—rabies, smallpox, yellow fever, scarlet fever, and typhus—are held to be caused by similar minute parasites, although these have not yet actually been seen and cultivated, but are surely inferred (from the nature and spread of these diseases) to exist.

The difference of the microbes called bacteria from the disease-causing microbes classed as "Protozoa" consists in their simpler structure and mode of growth. They are essentially filaments which continually multiply by fission—a process often carried so far that the little organisms present themselves as short rods, or as curved (comma-shaped), or even spherical particles (micrococci)—and only in favourable conditions arrest their self-division so as to grow for a time into the thread-like or filament shape. Often these filaments are not straight, but spirally twisted, and are called "spirilla." Some of them are blood parasites, but the larger number attack the tissues, and others occur in the digestive canal.

The parasitic disease-producing protozoa, on the other hand, are of softer substance, often have the habit of twisting themselves in a corkscrew-like manner, and usually are provided with an undulating membrane or frill, as well as with one or with two whip-like swimming processes (the latter are present also and are often numerous in the actively swimming phases of bacteria), and have a more complicated life-history. They divide, as a rule, longitudinally and not transversely, and pass from one "host" to a second, where they assume distinct forms—males and females, which conjugate and break up (each conjugated or fused pair) into a mass of very numerous, excessively minute, young. The disease-producing protozoa of this kind are frequently parasitic in the blood of man and animals, and were only recently recognised, after the disease-producing bacteria of many kinds had been thoroughly studied. These animal microbes are often spoken of as "blood-flagellates" or hæmo-flagellata, and the larger kinds are called "Trypanosomes," or "screw-form parasites," or whilst a series of more minute ones are called "Piroplasma," or "pear-shaped parasites." Many, but not all, are found during a certain period of their life, actually inside the corpuscles of the blood. The fact that many of these blood-flagellates (if not all) have, besides their life in the blood of one species of animal, a second period of existence in the juices or the gut of another animal, has made it very difficult to trace their migrations, since in the second phase of their history their appearance differs considerably from that which they presented in the first. And often they exist in one kind of animal without doing any harm, and are only poisonous when introduced by insects into the blood of other kinds of animals!

There is, further, another set of disease-causing protozoan parasites which are similar to the amœba or proteus-animalcule, and a third, which belong to the group of "ciliated infusoria." They are not so minute as the preceding set, and are not usually referred to as "microbes." They inhabit the intestine of man and animals, and cause, in some instances, dysentery. These two later kinds of protozoan parasites I will at the moment leave out of consideration, as well as the "coccidia," which multiply in the tissue-cells of animals—for instance, rabbits and mice—and cause an unhealthy growth and excessive multiplication of the cells of the tissues, which in some respects resembles that seen in the terrible disease known as cancer. Indeed, it is held by many investigators that some such parasite—though not yet discovered—is the cause of cancer.

A very important question is: How do these poison-producing parasites (for it is by the poison which they manufacture that they upset the healthy life of their hosts) make their way into the human body? The surface of the body of animals, like man, is protected by a delicate, horny covering—the epidermis—through which none of these parasites can make their way. They can only get through it, and so into the soft, juicy tissues and the fine blood-vessels which it covers, when it is cracked, broken, pierced, or cut. But they also have a way to open them through the softer moist surfaces of the inner passages, such as the digestive canal and the lungs. They enter (some kinds only and not a few) with food and drink into the digestive canal, and with the air into the air-passages and the lungs; and once in these chambers, which have only soft lining-surfaces, they are able to penetrate into the substance of the body. Many of those which enter the digestive canal do not require to penetrate further, but multiply excessively in the contents of the bowel, and there produce poisons, which are absorbed and produce deadly results—such are the bacteria which produce Indian cholera and ordinary diarrhœa—whilst the kind causing typhoid fever not only multiplies in the gut, but penetrates its surface.

The protective surface of man's body is broken, and the way laid open for the entrance of microbes in various ways. A slight scratch, abrasion, or even "chapping" is enough. Thus, a mere breaking of the skin of the knuckles by a fall on to dirty ground lets in the deadly bacterium of lock-jaw (tetanus), which is lurking in the soil. Leprosy is communicated from a leper in the same way. The almost ubiquitous bacteria of blood-poisoning (septicæmia) may enter by the smallest fissure of the skin, still more readily by large cuts or wounds. The bites and stabs of small and large animals—wolves, dogs, flies, gnats, fleas and bugs, also open the way, and often the deadly microbe has associated itself with the biting animal and is carried by it, ready to effect an entrance. Thus rabies (hydrophobia) is introduced by the bites of wolves and dogs, and a whole series of diseases, such as plague, malaria, sleeping-sickness, gaol-fever (typhus), yellow fever, relapsing fever, and others, are introduced into the human body by blood-sucking insects. Hence the immense importance of treating every slightest wound and scratch with chemicals (called "antiseptics"), which at once destroy the invading microbe—and of keeping a wounded surface covered and protected from their approach. In ways at one time unsuspected, such openings may be made by which poisonous microbes enter the body. Thus the little hard-skinned parasitic thread-worms which are often brought in by uncooked food into man's intestine, though by themselves comparatively harmless, scratch the soft lining of the bowel and enable poison-making microbes to enter the deeper tissues, and cause dangerous abscesses and appendicitis.

The carriers of disease germs thus become a very important subject of study. There are carriers which make no selection, but are, so to speak, "casual" in their proceedings, and there are others which have the most special and elaborate relations to some one kind of disease-causing microbe for which alone they are responsible, and to the life of which they are necessary. Let us look first at the more casual group. Man himself is a great carrier and distributor of his own diseases. Unless and until he has learned to be careful and guard against thoughtless proceedings, he is always spreading the microbes of his diseases and passing them on to his fellow men. He pollutes the waters, rivers, lakes, and pools from which others drink. He manures his crops, and then eats some of them uncooked. His hands are polluted by disease-causing microbes, and he handles (to an alarming and unnecessary extent) the food, such as bread and fruit, which is swallowed by his fellows, without cleansing it by heat. It has lately been shown that apparently healthy men and women often harbour within them the microbes of typhoid fever or of cholera (and probably other diseases), without themselves suffering in health, and that unsuspected they thus become distributing centres of these diseases. The names "typhoid carrier" and "cholera carrier" have actually been introduced to describe the condition of such persons. Then, again, by his breath, and by coughing and spitting, a man acts as a carrier to others of disease-microbes already lodged in him, as well as by actual contact in the case of those infections which are called "contagious." The numerous animals which surround and are associated with man act very largely as casual carriers and distributors of disease microbes. Thus dogs and even the cleanly cat are frequently carriers of disease. But more especially those creatures which visit man's food stores and food ready for consumption (such as bread, fruits, cold meat, etc.) are active carriers. Rats and mice run over such stores and pollute them. But the most widely active in this way is the common house-fly.

Whilst white men have developed an almost automatic resistance and objection to the visits of flies to their lips, eyelids, and any wound or scratch of the skin—a resistance which is not shown by many savage races—they yet allow house-flies to swarm in their dwellings, to run about and sample their food, with an indifference which is, when the truth is known, truly horrible in its fatuity and foolhardiness. For the fact is that the feet and proboscis of the common house-fly are covered with microbes of all sorts, picked up by his explorations upon every kind of filth. At every step which he takes he plants a few dozen microbes, which include those of infantile diarrhœa, typhoid, and other prevalent diseases. This is easily shown by allowing him to walk over a smooth plate of sterilised nutritive gelatine and preserving it afterwards free from the access of microbes from the air. In twenty-four hours every footstep of the fly on the gelatine is marked by an abundant and varied crop of microbes, which have multiplied from the individuals let drop by the little pedestrian. There is no doubt whatever that the house-fly is a main source of the dissemination of the microbe of infantile diarrhœa, and the cause annually of hundreds of thousands of deaths of children in the great cities of Europe and America. Also in camps and infected districts he is largely responsible for the introduction of the microbe of typhoid fever into the human food to which he has free access after his previous visits to open latrines. The house-fly is himself a product of dirt and neglect. The eggs are laid in old manure heaps and kitchen middens, and the maggots, which eventually are transformed into flies, nourish themselves in those accumulations. When this refuse is rapidly and regularly removed by the care of the sanitary officials of a town, the flies diminish in number, as they have diminished in London within the last thirty years. We no longer are overrun by flies in London in the summer months. The man selling sheets of sticky paper is no longer heard in our streets calling "Catch 'em alive, oh!" But in country places, where a neglected stable-yard is near the dining room of the inn, house-flies are as great a nuisance and danger as ever. There is no difficulty, if the simplest rules of cleanliness are observed, in abolishing them altogether from human association, but combined and simultaneous action against them is an essential condition of success.

IMMUNITY AND CURATIVE INOCULATIONS

During the last twenty years the whole attitude of the study and investigation of disease-causing microbes has advanced from the preliminary step of merely identifying certain microbes as the causes of certain diseases to a further step, viz. that of attempting to defend the animal and the human body against their attacks in the manner already so finely started by Pasteur. For many years disease after disease was examined and found to be caused by special bacteria or other microbes. Even non-infectious diseases or diseases only communicable under very special conditions were found to be due to microbes, so that it is probable that all disease that is not due to congenital malformation or to mechanical injury, or to poison fabricated in the weapons of larger animals and plants, or by man himself, is due to microbes. "Life," says Lord Justice Moulton, "is one ceaseless war against these enemies, and the periods of our too-transient successes are known as health." One of the last diseases traced to microbes is that sad condition known as "infantile paralysis," by which so many of the brightest and best members of the community have been crippled, from childhood onwards, through life.

Of late we have been making rapid strides in arriving at a knowledge as to how Nature herself protects higher creatures from the excesses and exuberance of destructive microbes, and we are now able to see that it is in adopting her methods that our best hope of increasing that protection lies. Nature is satisfied if the efficacy of her defence is sufficient to save enough individuals to carry on the race. Man desires in the case of his own fellows to out-do Nature and to save all.

A century and a half ago, before the true character of infective disease was understood, it was observed that an individual who was attacked by the smallpox and recovered became incapable of receiving the infection again.

He was "protected" or "immune." The practice of "inoculation" was introduced from the East by Lady Montague. The infectious matter was introduced from a smallpox patient into the person to be protected by rubbing it into a scarified part of the skin. A much less severe attack of smallpox was thus produced than that which usually followed the natural infection, which (though we do not know precisely its mode of entrance) is more widely spread through the blood. At the same time the condition of "immunity" after the attack was brought about with equal efficacy. When Jenner introduced inoculation with "cowpox" for the purpose of establishing "immunity" in the vaccinated person, inoculation with smallpox itself was a very usual practice. It was open to the objection that sometimes an unexpectedly violent attack of the disease was produced, resulting in death, and that the active infection was kept alive and ever present in the community. The notion with regard to the mode in which "immunity" was produced by either the Montacutian or Jennerian inoculation was, even after the general knowledge of microbes as the living contagion of disease had been arrived at, that the mild attack due to inoculation "used up" something in the blood—in fact, exhausted the soil, so that the infective matter or microbe could no longer flourish in the blood. And this view was accepted as the explanation of the "immunity" to the anthrax disease conferred on cattle and sheep by Pasteur's inoculations of weakened, but still actively growing, cultures of the anthrax bacillus. Another theory was that they produced something in the blood by their own life-processes which checked their further growth, just as yeast will not grow in wort in which it has produced 8 per cent. of alcohol, and as a fire may be choked by its own smoke or ashes.

We now know that both these explanations of "immunity" are incorrect. Nature provides at least three varieties of defence within the blood of higher animals against disease-producing microbes which have broken through the outer line of fortification, the skin. These three methods are effective in different cases (one in this disease, the other in that), and, on the whole, are sufficient to preserve the races of animals (including man) from complete destruction. These are (1) the production in the blood of an antidote to the toxin or poison elaborated by the invading microbe—an antitoxin, which chemically neutralises the toxin; (2) the production in the blood of the attacked animal of a "germicidal" poison which repels and kills the attacking microbes themselves (not merely neutralising their poisonous products); (3) the extermination of the intrusive, disease-producing microbes by a kind of police, which scour the blood channels and tissues and "eat up"—actually engulf and digest—the hostile intruders. These latter agents, actual particles of the living animal in which they exist, are the "eater-cells," or "phagocytes"—minute, viscid, actively moving cells, resembling the animalcules called "amœba." They are only the one two-thousandth of an inch in diameter, and are known as the white or colourless corpuscles of the blood. They are far less numerous than the red blood-corpuscles, which are the agents for carrying oxygen, but there are eight thousand million of them in a large spoonful of blood. They are the really important agents in protecting us from microbes, since they not only engulf and digest and so destroy those intruders, but it is probable (not certain) that they also are the manufacturers of the antitoxins and of the germicidal poisons.

If these three defensive processes given us by Nature are in working order, that is to say, if we are "healthy," they should secure to us a sufficient "immunity"—at at any rate, "recovery"—from any attack of disease-producing microbes. But they are not in "unselected," widely ranging mankind always equal (in their unaided natural state) to their task.

The attempts to produce immunity by vaccination with weakened or localised disease germs is really an attempt to train and develop to a high point the activities of the phagocytes or eater-cells of the blood.

The introduction of antitoxins by injection of them into the blood (as in the treatment of diphtheria, lock-jaw, and snake-bite) is an attempt to bring to the rescue of a patient who would sooner or later produce his own antitoxins (but perhaps too late or in insufficient quantity) the similar antitoxin obtained from the blood of another animal which has been artificially made to produce in its blood an excessive quantity of that substance.

Mithridates, King of Pontus, was, according to ancient legend, in consequence of his studies and experiments, soaked with all kinds of poisons to which he had become habituated by gradually increasing doses, and he had at last reached a condition in which no poison could harm him, so that when he was captured by the Romans and wished to kill himself (which was the correct thing in those days for a fallen king to do), he wept because he was unable to get any poisons which would act upon him. He was "immune" to all poisons. This real or supposed immunity resulting from the introduction into the living body at intervals of a series of doses of a poison gradually increasing strength has been called "Mithridatism," and animals and men so treated have been said to be "mithradatized." The toleration of poisonous drugs—such as tobacco and alcohol, and even of mineral poisons, such as arsenic—was, until lately, regarded as merely a special exhibition of that habituation of "adaptation by use" which living things often show in regard to some of the conditions of their life. Unusual cold, unusual heat, unusual moisture, salinity or the reverse, unusual deprivation of food, unusual muscular effort may be tolerated by animals without injury provided that they have been "gradually accustomed" to the unusual thing, or, in other words, that the unusual has been gradually made the usual; so that there is a saying that eels after a time even get used to being skinned. There was no attempt to explain the details of this process of habituation; it was assumed to be a part of the general "educability" of living matter.

The study of the education of living matter, in regard to various conditions which can act upon it, has yet to be further carried out, but the way in which the poisons made by disease germs and the like, and the disease germs themselves, are dealt with in the blood and tissues has, on account of its urgent importance, from a medical point of view, been already profoundly studied by experimental and microscopic methods of late years. The old notion as to "mithridatism" was that an animal or a man would have to be separately prepared and "immunised" by habituation for every distinct kind of poison. We now know that this is not the usual way in which Nature confers immunity to poisons. Most astonishing, and at first sight magical or mysterious, powers exist in the living protoplasmic cells in and around the blood of man and higher animals, which enable their possessors to resist and combat the poison-producing microbes, and also the poison itself, of all kinds, by which the race is liable to be attacked.

Few of us realise what a wonderful and exceptional fluid the blood of a higher animal is. The Australian natives attach so little importance to it that they actually cut themselves and use their blood as a sort of paste for sticking decorative feathers on to a pole! The Papuans are more advanced, since they regard the flow of blood from a cut or graze as an evil portent. And some respect to the greatness and wonder of blood is shown by those persons among civilised peoples (more frequently men than women) who faint when they see blood, or even at the mention of its name! This stream of red fluid within us (of which an average man has about fifteen pints in his vessels) courses at a tremendous rate from the heart through all the endless branches and networks of arteries, capillaries and veins, and back to the heart. It feeds, cleanses, warms and takes "vital air" (the old name for oxygen gas) dissolved in it to every particle of our bodies, fresh and fresh at every pulse-beat as it rushes on. It not only absorbs crude digested food through the walls of the gut, but conveys it to where it is worked up and distributes the worked-up product. It removes the quickly used-up substances from every part, and the choke-damp or carbonic acid which would stop the whole machine, and kill us, were it not got rid of through the lungs as the blood hurries through the walls of these air-sacs, whilst other used-up materials are carried by it to the kidneys and passed out of the body through them. Every part of the body is brought into common life with every other part by this impetuous blood-stream—which is here, there, and everywhere, right round, and back again, in twenty-five seconds! It is obviously a very serious thing if a poison-producing microbe gets into this blood-stream and multiplies within it, or if poison-producing microbes lodge somewhere beneath the skin in a wound, and keep on discharging virulent poison into the blood! The mischief is spread all over the body at once.

It is not surprising, then, that the long course of natural selection and survival of the fittest has resulted in the fixing in the blood and the living cells immediately connected with it of extraordinary protective powers. The floating scavenger cells (eater-cells or phagocytes, first recognised as such and so named by Metchnikoff) are already found in the blood of quite simple animals in worms, shell-fish and insects. I have watched them with the microscope at work in transparent minute living water-fleas eating up, and digesting microbes which had got into the water-flea's blood. In higher animals what we call "inflammation" is a condition—the result of a new and advantageous mechanism—which consists in a local retarding of the blood-current, effected by the action of the nerves on the muscular walls of the blood-vessels, and the consequent escape of the eater-cells into the injured or infected tissue, there to eat up and destroy the injurious microbes or other particles. Special and remarkable properties—chemical activities of an extraordinary character—have been gradually developed in the floating phagocytes and in similar non-floating fixed cells over which the blood flows.

These special chemical activities are of several distinct kinds. The first is the power to convert the poison of a microbe into a destroyer of that poison—toxin into antitoxin. The atoms of these poisons are elaborately composed combinations of the organic elements. By a "shake" or a "twist" (so to speak) administered by the living cells of the blood the combination is altered, and the toxin becomes an antitoxin, destroying by chemically combining with it the very toxin from which it was formed. This is a far more efficacious method than the supposed mithridatic "habituation" or "toleration" of a poison, with small doses of which you have to be gradually prepared. The healthy blood converts any one of a large series of microbe poisons into antitoxins. It is true that apparent "opposites" are often closely allied in Nature. Evil smells and tastes are closely allied to sweet perfumes and flavours, and what is healthy and agreeable to some men acts as virulent poison to others (e.g. shell-fish, egg, quinine, opium). The smallest change in the substance administered or the smallest difference in the living substance of an individual (what is called "idiosyncrasy") makes all the difference between "poison" and "meati."

If the phagocytes and similar cells in the blood of a man or animal exposed to the poison produced by localised microbes (such as those of tetanus, diphtheria and septic growths) cannot produce enough antitoxin so as to quickly destroy the poison, we can, and do, nowadays, save his life, by injecting into his blood the required antitoxin, obtained from another animal which we have caused (by injection of the toxin) to produce the antitoxin in excess. That is one sort of "immunity" or "resistance" which we can confer, and is largely in use at the present day—the "antitoxin" treatment.

The second poison-repelling chemical activity of the blood, produced by the living cells in and about it, consists in the blood becoming directly poisonous to injurious microbes. It becomes "bactericidal," produces a bactericidal poison (called an alexin) which is usually present in normal blood, but is greatly increased when large numbers of certain poisonous microbes (e.g. those of typhoid fever) get into the blood. Again, by other chemical substances produced in it, the blood may, without actually killing the invading bacteria, only paralyse them, and cause them to "agglutinate" (that is, to adhere to one another as an inactive "clot" or "lump"). As the "agglutinating" poison is peculiar (or nearly so) for each kind of microbe, we can tell whether a patient has typhoid by drawing a drop of his blood into a tube, and adding some fresh living typhoid bacilli to it. If the patient had typhoid he will have begun to form the "typhoid-agglutinating" or "typhoid-paralysing" poison in his blood, and the experiment will result in the "agglutination" (sticking together in a lump) of the typhoid bacilli. And so we prove, in a doubtful case, that the patient has typhoid.

The third chemical activity of the blood in dealing with poisonous microbes is also one which is conferred upon it by its living cells when excited by the presence of those microbes. It is the production of a "relish" (for so it must be called) which attaches itself to the microbes and renders them attractive to the eater-cells (the phagocytes), so that those swarming amœba-like floating particles at once proceed to engulf the microbes with avidity. In the absence of the relish (the Greek word for it used by Sir Almroth Wright, its discoverer, is "opsonin"), the eater-cells are sluggish—too sluggish—in their work. They resemble a child who will not eat dry toast, or, at best, only slowly, but will devour rapidly many pieces when the toast is buttered. It is of the utmost importance to us that our white corpuscles, or eater-cells, should not be sluggish but greedy.

There are some microbes which will produce deadly poison if grown in the clear fluid (serum) of the blood of an animal (as, for instance, the cholera-microbe when grown in the serum of the frog's blood), yet when inoculated living into the blood of that animal never cause the slightest illness! Why? Because they are at once eaten by the vigilant phagocytes of the blood before they can produce any appreciable amount of poison. That is easily demonstrated by experiment. Our main means of defence against microbial disease, says Metchnikoff—though cleanliness and precaution against access of microbes are all very well in their way—is the activity of our phagocytes. Now it appears that just as in the other cases I have been considering, so in the production of "relish," the power to produce it resides in the blood (and perhaps the cells of its vessels), but is not set at work until the enemy is in the blood. Suppose there is an infection, an invasion of the blood and tissues by one or other disease-causing microbe. Gradually if the body is healthy the "relish" is produced and becomes attached to the invading microbes. The phagocytes swallow them greedily and make an end of the invasion.

It is proved that this aroused avidity of the phagocytes is due to no change in the phagocytes themselves; since if they are transferred to the serum of a normal man they show no such predilection for the special invading microbe. The "opsonin," or "relish," is something exuded into or produced in the blood fluid when the attacking microbe arrives. It attaches itself to them: that is the essential fact. In many of us the phagocytes are not at a given moment so "avid" of this or that disease-microbe as they should be in order to protect us from its multiplication and poison production. But it is found that by injecting boiled and cooled (therefore dead) microbes of a particular kind into the blood of a man, you can start the production of the "relish" appropriate to that kind. The dead microbes answer this purpose; they excite the production of the opsonin appropriate to them and yet are not themselves dangerous, since they are dead. When subsequently (or possibly concurrently in small quantity) living microbes of the same disease enter the blood, the opsonin is ready for them. They are, to put it picturesquely, like oysters at the oyster-bar, peppered and vinegared "in no time," and then swallowed by the phagocytes by the dozen. This seems almost too comic a view of the deadly struggle of man and higher animals for health and freedom from the swarming pests which everywhere invade him. Yet it is correct, and involves a simple and fundamental truth. Our properties and appetites are but the sum of those of the protoplasmic organisms—the cells—of which we are built up. Our need for a relish with oysters is the same thing as the need of the phagocyte for a relish with its microbes, not something "poetically" compared to it. The story of "the oysters and the carpenter" might be replaced by that of "the microbes and the phagocyte." The saying, "Fine words butter no parsnips," finds a parallel in the remark that "The drinking of drugs does not opsonise microbes."

Half-way between us and the amœba-like unicellular organisms we find the earth-worm preparing his piece of lettuce (as Darwin showed) with a juice exuded from his mouth, a "relish" reminding one of the Kava drink of the South Sea Islanders. To "opsonise" or render attractive by the application of chemical "relish" is a proceeding which we find in operation in the feeding of the minute colourless corpuscles which engorge the still more minute bacteria—and also in the preparation of their food by various lower animals, and finally in the elaborate flavouring and cooking of his food by civilised man!

THE STRANGE STORY OF ANIMAL LIFE IN NEW ZEALAND

New Zealand consists of two islands, together more than 1,000 miles long and of about 200,000 square miles area. It is 1,000 miles distant from New Caledonia, the nearest island of any considerable size, and is 1,500 miles from the great Continental island of Australia. There is no other island in the world so large and at the same time so remote from other considerable tracts of land. Australia is closely connected by island groups at a distance of only 100 miles to Asia. The isolation of New Zealand is unique. The seas around it are of vast depth and of proportionately great age. During the chalk period—before the great deposits and changes of the earth's face which we assign to the Tertiary period—New Zealand consisted of a number of small scattered islands, which gradually, as the floor of the sea rose in that part of the world, became a continent stretching northward and joining New Guinea. In that very ancient time the land was covered with ferns and large trees. Birds (as we now know them) had only lately come into existence in the northern hemisphere, and when New Zealand for a time joined that area the birds, as well as a few lizards and one kind of frog, migrated south and colonised the new land. It is probable that the very peculiar lizard-like reptile of New Zealand—the "tuatara" or Sphenodon— entered its area at a still earlier stage of surface change. That creature (only 20 in. long) is the only living representative of very remarkable extinct reptiles which lived in the area which now is England, and, in fact, in all parts of the world, during the Triassic period, further behind the chalk in date than the chalk is behind our own day. For ages, this "type" with its peculiar beak-like jaws, has survived only in New Zealand. Living specimens have been brought to this country, and are to be seen at the

Zoological Gardens in Regent's Park. Having received, as it were, a small cargo of birds and reptiles, but no hairy, warm-blooded quadruped, no mammal, New Zealand became at the end of the chalk-period detached from the northern continent, and isolated, and has remained so ever since. Migratory birds from the north visited it, and at a late date two kinds of bat reached it and established themselves.

Thus we are prepared for the very curious state of things in this large tract of land. Looking at New Zealand as it was a thousand years ago, we find there were no mammals living on it excepting a couple of bats and the seals (so-called sea lions, sea elephants, and others) which frequent its coasts. There were 180 species of birds, and many of these quite peculiar to the island. Many of the birds showed in the absence of any predatory enemies—there being no carnivorous quadrupeds to hunt them or their young—a tendency to lose the power of flight, and some had done so altogether. The gigantic, wingless Moas—allied to the ostrich and the cassawary—had grown up there, and were the masters of the situation. There were many species of these—one of great height—one fourth taller than the biggest known ostrich; others with short legs of monstrous thickness and strength. Allied to these are the four species of Kiwi or apteryx, still existing there. They are very strange wingless birds, about the size of a large Dorking fowl. The Kiwis are still in existence, but the Moas and some of the other flightless birds have died out since the arrival of the Maori man, who killed and ate them.

A bird which was believed sixty years ago both by the natives and white men to have become extinct, the Takahe, or Notornis, was known by its bones and from the traditions of the natives. Much to the delight of naturalists, four live specimens of it were obtained at intervals in the last century, the last as late as 1898. The beautiful dark plumage and thick and short beak, which is bright red, as are the legs, are well known from the two specimens preserved in the Natural History Museum. The Notornis is a heavy, flightless "rail." Rails are remarkable for their size and variety in New Zealand, where there are twenty species, some of them very sluggish in flight, or like Notornis, flightless (the wood hens). Amongst the flightless birds of New Zealand is a duck, as helpless as the heaviest farmyard product, and yet a wild bird, and then there are the penguins, which swim with their wings, but never fly, and belong entirely to the southern hemisphere. Many species are found on the shores of New Zealand. Other noteworthy birds of New Zealand are the twelve kinds of cormorants, the wry-bill plover, the only bird in the world with its beak turned to one side, the practically flightless Kakapo, or ground parrot (Stringops), the Huia, a bird like a crow in appearance, whose male has a short straight beak, whilst the female has a long one, greatly curved; the detested Kea, the parrot which kills the sheep, introduced by the colonists, by digging out with its beak from their backs the fat round the kidneys; also very peculiar owls and wrens, and the fine singing bell-birds.

The peculiarity of the indigenous animals of New Zealand is seen not only in the absence of mammals and the abundance of remarkable birds, many of them flightless, but also in the fact that there are no snakes in this vast area—no crocodiles, no tortoises—only fourteen small kinds of lizard (seven Geckoes and seven Skinks), and only one species of frog (and that only ever seen by a very few persons)! There were fish in the rivers when settlers arrived there, but none very remarkable. Insects and flies of every kind, scorpions, spiders, centipedes, land-snails and earth-worms were all flourishing in the forests of New Zealand a thousand years ago, serving in large measure as the food of birds, fish and lizards. The great island continent of Australia, 1,500 miles away, is peculiar enough in its living products, quite unlike the rest of the world in its egg-laying duck-mole and spiny ant-eater, and in its abundant and varied population of pounched mammals or marsupials, emphasized by the absence (except for two or three peculiar little mice and the late-arrived black-fellow and bush-dog) of the regular type called "placental" mammals which inhabit the rest of the world. The rest of the world except New Zealand! Strange as Australia is, New Zealand is yet stranger. Long as the isolation of Australia has endured, and archaic and primitive in essential characters as is its living freight of animals and plants navigated (as it were) in safety and isolation to our present days, yet New Zealand has a still more primitive, a more ancient cargo. When we divide the land surfaces of the earth according to their history as indicated by the nature of their living fauna and flora and their geological structure, and the fossilised remains of their past inhabitants, it becomes necessary to separate the whole land surface into two primary sections: (a) New Zealand, and (b) the rest of the world, "Theriogœa," or the land of beasts (mammals). Then we divide Theriogœa into (1) the land of Marsupials (Australia) and (2) the land of Placentals (the rest of the world). This last great area is divisible according to the same principles into the great northern belt of land, the Holarctic region and the (three not equally distinct) great southward-reaching land surfaces—the Neo-tropical (South America), the Ethiopian (Africa, south of the Sahara), and the Oriental (India and Malay).

The bird-ruled quietude of New Zealand was disturbed 500 years ago by the arrival of the Polynesian Islanders, the Maoris, in their canoes. They brought with them three kinds of vegetables which they cultivated, a dog and a kind of rat. The dogs soon died out, but the rat has remained, and is considered to have done little or no harm. It was not one of the destructive proliferous rats of the northern hemisphere. The Maoris hunted the big birds—the Moas and others—for their flesh, and ate their eggs, and it is probable that they caused or accelerated the extinction of the Moa and two or three other birds. In the north island they nearly exterminated the white heron, the plumes being valued by them. On the whole, very little damage was done to the natural products of the islands by the Maoris. "It was with the advent of the Europeans," says Mr. John Drummond, F.L.S., in his interesting and well-illustrated book on 'The Animals of New Zealand,' "that destruction began in earnest. It seemed as if they had been commanded to destroy the ancient inhabitants." They killed right and left, and, in addition, burnt up the primæval forests and bushes till a great part of the flora was consumed. It was never a very varied or strong one, consisting only of some 1,400 species, which are now in large proportion vanishing, whilst 600 species of plants, most of them introduced accidentally rather than intentionally by the European settlers, have taken their place.

Here I may state the great principle which, in regard to plants as well as animals, determines the survival of intruders from one region to another. It appears that setting aside any very special and peculiar adaptations to quite exceptional conditions in a given area, the living things, whether plants or animals, which are brought to or naturally arrive at such an area, survive and supplant the indigenous plants and animals of that area, if they themselves are kinds (species) produced or formed in a larger or more variegated area; that is to say, formed under severer conditions of competition and of struggle with a larger variety of competitors, enemies and adverse circumstances in general. Thus, the plants of remote oceanic islands are destroyed, and their place and their food are taken by the more hardy "capable" plants of Continental origin. And, in accordance with the same principle, as Darwin especially maintained, the plants of the northern hemisphere, produced as they are in a wide stretching belt of land— Europe, temperate Asia, and North America—always push their way down the great southern stretches of land (by cool mountain roadways), and when they have arrived in the temperate regions of the southern hemisphere, they have at various geological epochs starved out, taken the place of, or literally "supplanted" the native southern flora, which in every case has been formed on a narrow, restricted and peninsula-like area. The same greater "potency" of the animals of the Holartic region has in the past established them as intruders into South America, Ethiopia and India, and has led to the inevitable survival of the animal of the large area when brought into contact with the animal of the small and restricted area. Applying these principles to New Zealand, we see that no country, no area of land, could have a worse chance for the survival of its animal and vegetable children than that mysterious land, isolated for many millions of years in the ocean, the home of the Tuatara, solitary survivor of an immensely remote geologic age, the undisturbed kingdom of huge birds, so easy-going that they have ceased to fly, and have even lost their wings!

The first European animals to settle there were the pigs benevolently introduced into New Zealand by Captain Cook. They multiplied apace, served for food and sport both to the natives and the early settlers, and destroyed the ancient Triassic reptile, the Tuatara, which only survives now on rocky islands near the coast. In less than a hundred years the settlers had introduced sheep and cattle, and looked upon the abounding pigs as a scourge. In 1862, pig-hunters were employed to destroy them—three hunters would kill 20,000 pigs in a year. Dogs, cats and the European rats came in early with the settlers, and destroyed the flightless birds, driving them for shelter to the mountains. As the settlers increased they shot down millions of birds of all kinds, and burnt up grass, shrub, and bush. At last, a few years ago, the Government established three islands as "sanctuaries," where many of the more interesting birds survive, and are increasing.

Besides cattle and sheep (which have flourished exceedingly) the colonists introduced rabbits, pheasants and the honey-bee, and later on quails, hares, deer, and trout. Clover depends on bees for its fertilisation and seeding. White clover, taken over there for pasture, did not seed in New Zealand until the honey-bee was imported in 1842, and later, as they could not seed red-clover without it, the colonists had to introduce the humble-bee, and the red-clover now also seeds freely and the imported farm-beasts have their accustomed food. Besides the animals already named, the colonists have introduced ferrets and weasels, to reduce the destructive excess of the imported rabbits; and they, whilst failing to subdue the rabbits, have themselves become a serious nuisance. Of small birds there were introduced the house-sparrow, which is too prolific, and is hated by the farmers; the greenfinch, a pest; the bullfinch, a failure. The introduced skylark and the blackbird (alas! poor colonists) are not the joy of New Zealanders—the farmers hate them. The European settlers had the audacity to introduce also the most beautiful and beloved of all birds, our own perfect "Robin Redbreast," and they add want of manners to their violent and uncalled-for hospitality by speaking ill of this sweetest and brightest of living things. After this, I am rather glad to report that the esteemed table-delicacies, pheasants and partridges, don't get on well in New Zealand; nor do turtle-doves. The thrush is spreading and meets with the approval of the hypercritical New Zealander. The hedge-sparrow, the chaffinch and the goldfinch have flourished abundantly, but the linnet has failed. A very interesting and important problem for New Zealand naturalists to solve is that as to why one bird succeeds in their remote land and another does not. The British trout have grown to an enormous size and are destroying all other fresh-water life. Imported red-deer flourish, and are shot with great satisfaction by the colonists. The American elk has been introduced in the South Island, and the mountain goats—the ibex and the thar—are to be acclimatized in the mountains, so that unnatural sport may flourish in this ancient land of quiet and of wondrous birds, turned topsy-turvy by enlightened man.

THE EFFACEMENT OF NATURE BY MAN

Very few people have any idea of the extent to which man since his upgrowth in the late Tertiary period of the geologists—perhaps a million years ago—has actively modified the face of Nature, the vast herds of animals he has destroyed, the forests he has burnt up, the deserts he has produced, and the rivers he has polluted. It is, no doubt, true that changes proceeded, and are proceeding, in the form of the earth's face and in its climate without man having anything to say in the matter. Changes in climate and in the connections of islands and continents across great seas and oceans have gone on, and are going on, and in consequence endless kinds of animals and plants have been, some extinguished, some forced to migrate to new areas, many slowly modified in shape, size, and character, and abundantly produced. But over and above these slow irresistible changes there has been a vast destruction and defacement of the living world by the uncalculating reckless procedure of both savage and civilised man which is little short of appalling, and is all the more ghastly in that the results have been very rapidly brought about, that no compensatory production of new life, except that of man himself and his distorted "breeds" of domesticated animals, has accompanied the destruction of formerly flourishing creatures, and that, so far as we can see, if man continues to act in the reckless way which has characterised his behaviour hitherto, he will multiply to such an enormous extent that only a few kinds of animals and plants which serve him for food and fuel will be left on the face of the globe. It is not improbable that even these will eventually disappear, and man will be indeed monarch of all he surveys. He will have converted the gracious earth, once teeming with innumerable, incomparably beautiful varieties of life, into a desert—or, at best, a vast agricultural domain abandoned to the production of food-stuffs for the hungry millions which, like maggots consuming a carcase, or the irrepressible swarms of the locust, incessantly devour and multiply.

Another glacial period or an overwhelming catastrophe of cosmic origin may fortunately, at some distant epoch, check the blind process of destruction of natural things and the insane pullulation of humanity. But there are, it seems probable, many centuries of what would seem to the men of to-day deplorable ugliness and cramping pressure in store for posterity unless an unforeseen awakening of the human race to the inevitable results of its present recklessness should occur. Whatever may be the ultimate fate of the earth under man's operations, we should endeavour at this moment to delay, as far as possible, the hateful consummation looming ahead of us.

It is interesting to note a few instances of man's destructive action. Even in prehistoric times it is probable that man, by hunting the mammoth—the great hairy elephant—assisted in its extinction, if he did not actually bring it about. At a remote prehistoric period the horses of various kinds which abounded in North and South America rapidly and suddenly became extinct. It has been suggested, with some show of probability, that a previously unknown epidemic disease due to a parasitic organism—such as those which we now see ravaging the herds of South Africa—found its way to the American continent. And it is quite possible that this was brought from the other hemisphere by the first men who crossed the Pacific and populated North America.

To come to matters of certainty and not of speculation, we know that man by clearing the land, as well as by actively hunting and killing it, made an end of the great wild ox of Europe, the aurochs or urus of Cæsar, the last of which was killed near Warsaw in 1627. He similarly destroyed the bison, first in Europe and then (in our own days) in North America. A few hundred, carefully guarded, are all that remain in the two continents. He has very nearly made an end of the elk in Europe, and will soon do so completely in America. The wolf and the beaver were destroyed in these British Islands about 400 years ago. They are rapidly disappearing from France, and will soon be exterminated in Scandinavia and Russia and in Canada. At a remote prehistoric period the bear was exterminated by man in Britain and the lion driven from the whole of Europe, except Macedonia, where it still flourished in the days of the ancient Greeks. It was common in Asia Minor a few centuries ago. The giraffe and the elephant have departed from South Africa before the encroachments of civilised man. The day is not distant when they will cease to exist in the wild state in any part of Africa, and with them are vanishing many splendid antelopes. Even our "nearest and dearest" relatives in the animal world, the gorilla, the chimpanzee and the ourang, are doomed. Now that man has learnt to defy malaria and other fevers the tropical forest will be occupied by the greedy civilised horde of humanity, and there will be no room for the most interesting and wonderful of all animals, the man-like apes, unless (as we may hope in their case, at any rate) such living monuments of human history are made sacred and treated with greater care than are our ancient monuments in stone. Smaller creatures, birds like the dodo and the great auk and a whole troop of others less familiar, have disappeared and are disappearing under the human blight. Even some beautiful insects—the great copper butterfly and the swallow-tail butterfly—have been exterminated in England by human "progress" in the shape of the drainage of the Fen country.

But the most repulsive of the destructive results of human expansion is the poisoning of rivers, and the consequent extinction in them of fish and of well-nigh every living thing, save mould and putrefactive bacteria. In the Thames it will soon be a hundred years since man, by his filthy proceedings, banished the glorious salmon, and murdered the innocents of the eel-fare. Even at its foulest time, however, the Thames mud was blood-red (really "blood-red," since the colour was due to the same blood-crystals which colour our own blood) with the swarms of a delicate little worm like the earth-worm, which has an exceptional power of living in foul water, and nourishing itself upon putrid mud. In old days I have stood on Hungerford Suspension Bridge and seen the mud-banks as a great red band of colour, stretching for a mile along the picture when the tide was low. In smaller streams, especially in the mining and manufacturing districts of England, progressive money-making man has converted the most beautiful things of nature—trout streams—into absolutely dead corrosive chemical sewers. The sight of one of these death-stricken black filth-gutters makes one shudder as the picture rises, in one's mind, of a world in which all the rivers and the waters of the sea-shore will be thus dedicated to acrid sterility, and the meadows and hill-sides will be drenched with nauseating chemical manures. Such a state of things is possibly in store for future generations of men! It is not "science" that will be to blame for these horrors, but should they come about they will be due to the reckless greed and the mere insect-like increase of humanity.

In the destruction of trees and all kinds of plants man has deliberately done more mischief than in the extermination of animals. By inadvertence he has completely abolished the strange and remarkable trees and shrubs of islands—such as St. Helena—where the herbivorous animals introduced by him have made short work of the wonderful native plants isolated for ages, and have completely exterminated them, so that they are "extinct." We have just had the opportunity of studying one of the few oceanic islands—"Christmas Island" (forty square miles in area)—untouched by man until thirty years ago. It lies 200 miles south of Java. Its native inhabitants, plants and animals were carefully examined, and specimens secured twenty years ago. There were then no human inhabitants, and the island was rarely visited. It was, however, about twelve years ago handed over by its proprietors to some thousand Chinamen to dig and ship the 15,000,000 tons of valuable "phosphate" (at a profit of a guinea a ton), which forms a large part of its surface. And now from time to time we shall have reports of this result of contact with man, and through him with all the plagues and curses of the great world. Already a remarkable shrew-mouse and two native species of rat, peculiar to the island, have disappeared. Dr. Andrews ("Proceedings of the Zoological Society," February 2nd, 1909), who has twice explored the island, gives evidence that this is caused by a parasitic disease (due to a trypanosome like those which cause sleeping-sickness and various horse and cattle diseases) introduced by the common black rats from the ships which now frequent the island. The further progress of destruction will be carefully and minutely observed and recorded—but not arrested!

It is, however, in cutting down and burning forests of large trees that man has done the most harm to himself and the other living occupants of many regions of the earth's surface. We can trace these evil results from more recent examples back into the remote past. The water supply of the town of Plymouth was assured by Drake, who brought water in a channel from Dartmoor. But the cutting down of the trees has now rendered the great wet sponge of the Dartmoor region, from which the water was drawn all the year, no longer a sponge. It no longer "holds" the water of the rainfall, but in consequence of the removal of the forest and the digging of ditches the water quickly runs off the moor, and subsequently the whole countryside suffers from drought. This sort of thing has occurred wherever man has been sufficiently civilised and enterprising to commit the folly of destroying forests. Forests have an immense effect on climate, causing humidity of both the air and the soil, and give rise to moderate and persistent instead of torrential streams. Spain has been irretrievably injured by the cutting down of her forests in the course of a few hundred years. The same thing is going on, to a disastrous extent, in parts of the United States. Whole provinces of the Thibetan borders of China have been converted into uninhabitable, sandy desert, where centuries ago were fertile and well-watered pastures supporting rich cities, in consequence of the reckless destruction of forest. In fact, whether it is due to man's improvident action or to natural climatic change, it appears that the formation of "desert" is due in the first place to the destruction of forest, the consequent formation of a barren, sandy area, and the subsequent spreading of what we may call the "disease" or "desert ulcer," by the blowing of the fatally exposed sand and the gradual extension, owing to the action of the sand itself, of the area of destroyed vegetation. Sand-deserts are not, as used to be supposed, sea-bottoms from which the water has retreated, but areas of destruction of vegetation—often (though not always) both in Central Asia and in North Africa (Egypt, etc.), started by the deliberate destruction of forest by man, who has either by artificial drainage starved the forest, or by the simple use of the axe and fire cleared it away.

The great art of irrigation was studied and used with splendid success by the ancient nations of the near East. They converted deserts into gardens, and their work was an act of compensation and restitution to be set off against the destructive operations of more barbarous men. But they, too, long ago were themselves destroyed by conquering hordes of more ignorant but more war-like men, and their irrigation works and the whole art of irrigation perished with them. One of the absolutely necessary works to be carried out by civilised man, when he has ceased to build engines of war and destruction, is the irrigation of the great waterless territories of the globe. A little home-work of the kind has been carried on in Italy regularly year by year since the days of Leonardo da Vinci, and our Indian Government is slowly copying the Italian example. In Egypt we have built the great dam of Assouan, whilst in Mesopotamia it is proposed to re-establish the irrigation system by which it once was made rich and fertile. But, as has lately been maintained by Mr. Rose Smith in his book, "The Growth of Nations," the vast possibilities of irrigation have not yet been realised by the business men of the modern world. Millions of acres in the warmer regions of the earth now unproductive can be made to yield food to mankind and rich pecuniary profits to the capitalists who shall introduce modern engineering methods and a scientific system of irrigation into those areas.

The whole problem of the increase of the more civilised races and the necessary accompanying increase of food-production depends for its solution on the speedy introduction of irrigation methods into what are now the great unproductive deserts of the world.

THE EXTINCTION OF THE BISON AND OF WHALES

The almost complete and very sudden disappearance of the bison in North America thirty years ago does not seem to have been due simply to the slaughter of tens of thousands of these creatures by men who made a commerce of so-called "buffalo-rugs." These "hunters" miscalled the unhappy bison, which is not a buffalo, nor at all like that creature, just as they gave the name "elk" to the great red deer (the wapiti), although there was a real elk, the so-called "moose," staring them in the face. The sudden extinction of the bison resulted partly from the slaughter and partly from the breaking up of the herds and the interference with their free migration by the trans-continental railway. An interesting discovery made only this year, in regard to the closely allied European bison, suggests that disease may also have played a part in the destruction of the North American bison. A few hundred individuals of the European bison are all that remain at this day. Some are carefully preserved by the Emperor of Russia in a tract of suitable country in Lithuania and another herd exists in the Caucasus. Some of the Lithuanian bison have lately been dying in an unaccountable way, and on investigating a dead individual a Russian observer has discovered a "trypanosome" parasite in the blood. The trypanosomes are microscopic corkscrew-like creatures, of which many kinds have become known within the last ten or fifteen years. They are "single cells"—that is to say, "protoplasmic" animalcules of the simplest structure—provided with a vibrating crest and tail by means of which they swim with incessant screw-like movement through the blood. They rarely exceed one thousandth of an inch in length exclusive of the tail. The poisons which they produce by their life in the blood are the cause of the sleeping-sickness of man (in tropical

Africa), of the horse and cattle disease carried by the tsetze fly, and of many similar deadly diseases—a separate "species" being discovered in each disease. A peculiar species is found in the blood of the common frog, and another in that of the sewer-rat. The last discovery of a "trypanosome" is that of one in the blood of the African elephant, announced to the Royal Society by Sir David Bruce.

It is a matter of great interest that a trypanosome has been found in a death-stricken herd of European bison. It suggests that one of the causes of the disappearance of the bison, both in Europe and America, may be the infection of their blood by trypanosomes, and that possibly, whilst a freely migrating and vigorous herd would not be extensively infected, a dwindled and confined herd may be more liable to infection, and that thus the final destruction of an already decadent animal may be brought about. It would now be a matter of extreme interest to ascertain whether the few dwindled herds of bison in North America are infected by trypanosomes, and no doubt we shall soon receive reports on the subject.

A most interesting branch of this subject of the unthinking extermination of great animals by man is that of the extermination of whales. Man is worrying them out of existence. Some are already beyond saving. It would be interesting to know whether there are trypanosomes or other blood-parasites in whales. I suppose that no one has an ill-feeling towards whales. Most of us have never seen a whale, either alive or in the flesh—only a skeleton. I have seen a live whale or two off the coast of Norway; and I once, in conjunction with my friend Moseley, when we were students at Oxford, cut up one, 18 ft. long, which had been exhibited for three weeks during the summer in a tent on the shores of the Bristol Channel, where we purchased it. The skeleton of that whale is now in the museum at Oxford, but happily the smell of it exists only in my memory. The late Mr. Gould, who produced such beautifully illustrated books on birds, told me that he once fell into the heart of a full-sized whale, which he was cutting up. He narrowly escaped drowning in the blood. The whale was not very fresh, and Mr. Gould was unapproachable for a week.

An immense number of whales are killed every year for their oil, and their highly nutritious flesh is wasted. There was an attempt some years ago to make meat extract from it. Some which was brought to me reminded me of the whale on the shores of the Bristol Channel. I do not know if the extract has proved palatable to other people. The Norwegians are specially expert in killing whales. They have been allowed to set up "factories" on the west coast of Ireland and in the Shetlands, where they kill whales with harpoons fired from guns, cut them up, and boil down the fat.

Whales are warm-blooded creatures which suckle their young, and have been developed in past geological times from land animals—the primitive carnivora—which were also the ancestors of dogs, bears, seals and cats.

Whales have lost the hind limbs altogether and developed the forelegs into fingerless flippers, whilst the tail is provided with "flukes" like the fins of a fish's tail in shape, but horizontal instead of vertical. The whole form is fish-like, the skin smooth and hairless. It is a remarkable conclusion arrived at by the investigators of the remains of extinct animals that a little four-legged creature the size of a spaniel, and intermediate in character between a hedgehog and a dog, was the common ancestor from which have been derived such widely different creatures as the whale and the bat, the elephant and the man. We can at the present day trace with some certainty the gradual modifications of form by which in the course of many millions of years the change from the primitive, dog-like hedgehog to each of those four living "types" has proceeded.

The whales of to-day are divided into the toothed whales and the whalebone whales. The great cachalot or sperm whale is captured, chiefly in the Southern Ocean, and killed in large numbers for the sake of the "spermaceti," or "sperm oil," which forms the great mass of its head, but he is so fierce and active that he is not easily captured, and is not in immediate danger of extinction. The smaller toothed whales, the killers, dolphins, and porpoises (though one of them—the bottle-nosed whale—is being killed out), are not as yet seriously threatened by commercial man. But the whalebone whales are in a parlous state. The Right whales, as they are called, are the chief of these. They are huge creatures, 60 ft. in length, with an enormous head: it is as much as one third of the total length in the Greenland whale. Besides the Greenland species there are four other "right whales," which may be considered as four varieties of one species. The head is not quite so large in them. The Biscay whale is one of them, and was hunted until it was exterminated in the Bay of Biscay, when the whalers, extending their operations further and further north, came upon the Greenland whale, which proved to be even more valuable than the Biscay species. The huge mouth in these two whales has hanging from its sides within the lips a series of long bars or planks of wonderfully strong, elastic, horny substance—the "baleen" or "whalebone"—each plank being as much as eight or in rare cases twelve feet long. Following close on one another and having hairy edges, they act as strainers so as to separate the floating food of the whale from the water which rushes through its mouth as it swims. The whalebone is of great value commercially, as is also the fat or oil. A hundred years ago whalebone fetched only £25 a ton, now the same quantity fetches more than £1,500. The Rorquals, or "Finners," have smaller heads and mouths; their whalebone is so short as to be valueless, but they grow to even greater size than the Right whales and are found on our own coasts and all over the world. The Humpback whale is one of these "Finners," distinguished by its excessively long flippers and huge bulk. The Biscay whale was the first of these great creatures to be hunted. The

Basques began its capture as early as the ninth century. It was exterminated by them in the Bay of Biscay, and only saved from complete extinction elsewhere by the discovery of the more valuable Arctic or Greenland whale. The capture of the Greenland whale began in 1612; and in 200 years the unceasing pursuit of this species had driven it to the remote places of the Arctic Ocean. It is now so rare that it is not worth while to send a ship out for the purpose of hunting it, and it will probably never recover its numbers. An idea of its value and former abundance may be formed from the fact that between 1669 and 1778 it yielded to 1,400 Dutch vessels about 57,000 individuals, of which the baleen and oil produced a money value of four million pounds sterling. Of late years a single large Greenland whale would bring £900 for its whalebone and £300 for its oil. These two great Right whales having been practically exterminated, the merciless hunt has now been turned on to the wilder and less valuable Finback whales or Finners. In these days of steam and electric light the Arctic night is robbed of its terrors, and the whale chase goes on very fast. The shot harpoon was invented in 1870 by Sven Foyn, a Norwegian, and is the most deadly and extraordinary weapon ever devised by man for the pursuit of helpless animals. It is this invention (a commercial, not a scientific, discovery!) which has, in conjunction with swift steamships, rendered the destruction of whales a matter of ease and deadly certainty. It is this which is being used on the Irish as on the Scandinavian coast, resulting in the pollution of the air and water by the carcases of the slaughtered beasts from which the oil has been extracted. This revolting butchery, without foresight or intelligence, is carried on solely for the satisfaction of human greed, and apparently will be stopped only by the extinction of the yet remaining whales. In forty years in the middle of last century the whale fishery of the United States yielded 300,000 whales to 20,000 voyages, and a value of sixty-million pounds sterling in baleen and oil. It is calculated that in the thousand years during which man has hunted the great whales not less than a million individuals have been captured. Man's skill and capacity have now become such that he will soon have cleared the ocean of these wonderful creatures, since, like the bison, the whales cannot persist when harried and interfered with beyond a certain limited degree.

It appears that the curious musk ox, which now lives on the fringe of the Arctic circle, and in the glacial period existed in the Thames Valley, is doomed. There (as in similar instances in other lands), the comparatively harmless savage race of men (in this case the Eskimo), whose weapons did not enable them seriously to threaten the existence of the animals around them, have now obtained efficient firearms. The musk ox is consequently now between two lines of fire—that of the white hunter on the south, and of the Eskimo on the north.

From regions far remote from the Arctic complaints come of an even more reckless destruction of helpless animals. Perhaps our legislators may feel some personal concern in this case, since it is neither more nor less than the approaching extinction of the turtle, the true green turtle of City fame, to eat which at the invitation of City dignitaries is one of the few duties of a legislator. Both the green turtles and the tortoise-shell turtles are being destroyed indiscriminately on the coast of Florida and in many West Indian Islands by brutal, careless, "white" beach-combers and idlers. By proper care of the eggs and young the turtles could easily be increased enormously in number, and a regulated capture of them be made to yield a legitimate profit. But neither the United States Government nor our own take any steps to restrain promiscuous slaughter of the turtles which come to the shore in order to lay their eggs. Soon the City Fathers will have to do without the "green fat" and their wives without tortoise-shell combs. It will serve them right. Such destitution in these—and, be it noted, in many other matters—will deservedly fall upon those who ignorantly, wilfully, and contentedly neglect to take steps to understand and to control the withering blight created by modern man wherever he sets his foot.

MORE ABOUT WHALES

The possibility of protecting whales from wanton slaughter by man is, no doubt, a matter open to discussion. Protection has, however, been accorded to one particular whale in an exceptional instance. Passenger steamers along the coast of New Zealand used to call at a station in a narrow inlet of the coast, called Pelorus Sound. A black whale, said to be of the kind known as Risso's Grampus, of about 14 ft. in length, was apparently a settled inhabitant of this channel, and used to follow the steamers and accompany them through the sound. He became famous and popular, and was known as "Pelorus Jack." He was always looked for and recognised by the sailors and passengers. Certain savagely destructive persons on one of these steamers—to the horror and disgust of the New Zealand world—made an attempt to shoot "Pelorus Jack." It is stated, and believed by sailors, that ill-luck consequently fell on that steamer. On its next voyage it was avoided by the whale, who had never failed to welcome friendly and non-aggressive steamships, and on a third voyage the steamer was wrecked. The feeling about "Pelorus Jack" was so strong that his Excellency the Governor of New Zealand, Lord Plunket, signed, on September 26th, 1904, an Order in Council, protecting "Pelorus Jack" by name for five years, and any person interfering with him was made liable to a fine of £100.

It appears that under the New Zealand Sea Fisheries Act of 1894 the Governor in Council is empowered to make regulations protecting any fish. Although zoologically not belonging to the class of fishes, whales are, technically and for all legal and commercial purposes "fishes," since they are "fished" and are the booty of "fisheries." I believe that no Governor, Council, or Secretary of State has power in the British Islands similar to that conferred on the Governor of New Zealand by a modern State which desires good and effective government. Such power is needed in all parts of the British Empire.

The whales, as compared with their dog-like ancestors, are modified to a more extreme degree and in more special ways than is the case in any other group of which we can trace the history over a similar period of development. This is connected with the complete change of conditions of life to which these mammals ("warm-blooded, air-breathing quadrupeds which suckle their young") have become adapted in passing from a terrestrial to a marine existence. Other mammalian ancestors have independently taken to a marine life and given rise to strange-looking adaptations, namely, the seals and also the Manatee and Dugong known as the Sirenians (so-called because they give rise to sailors' stories of mermaids and sirens), but these are far less changed, less modified than the whales. The whales have acquired a completely fish-like form. They frequently have a large back fin, and have lost the hind legs altogether. The horizontally spread flukes of the whale's tail have nothing to do with the hind legs, whereas the common seal's hind legs are tied together so as to form a sort of tail. In the bigger whales, sunk deep in the muscle and blubber, we find on each side well forward in the body (not near the tail) a pair of isolated, unattached bony pieces, which are the hip-bone and thigh-bone—all that remains of the hind limbs. The neck is so short that in many whales the seven neck-bones, or "vertebræ," are all fused into one solid piece not longer than a single ordinary vertebra, and showing six grooves marking off the seven vertebræ which have united into one.

The head is more strangely altered than any other part of the whale. The jaws are greatly elongated—so as to give a beak-like form in all—but this region is specially long and narrow in the "beaked whales" known to zoologists by the name Ziphius, in which it consists of a solid piece of ivory-like bone, which we find in a fossil state in the bone-bed of the Suffolk Crag. Farther back the bones of the face are suddenly widened in all whales and porpoises, and in many these bones grow up into enormous crests and ridges. The nostrils, instead of being placed, as in other animals, at the free end of the snout or beak, lie far back, so as to form the "blow-hole," which is near the middle of the head.

The circulation of the blood and the breathing of whales (including in that term the smaller kinds known as dolphins and porpoises) is still a matter which is not properly understood. When a Greenland whale is struck by the harpoon it dives vertically downward to a depth of 400 fathoms and more (nearly half a mile), and occasionally wounds the skin and bones of its snout by violently striking it on the sea-bottom. It remains below as long as forty minutes. Physiologists wish to know how the sudden compression of the air in the lungs in plunging to this depth and the equally sudden expansion of it in rising from such a depth is dealt with in the whale's economy, so as to prevent the absolutely deadly results which would ensue were any ordinary air-breathing animal subjected to such changes of pressure. Man can endure without suffering an increase of pressure of the gases in his body amounting to three or four times that to which he is accustomed, as, for instance, when working in the compressed air of "caissons." But the whale goes suddenly to a depth at which the pressure is eighty times that at the surface! Then, too, man (and other terrestrial animals), after being subjected (for instance, in a caisson) to a pressure of four times that which exists on the free surface of the earth, is liable to be killed by suddenly passing from that high pressure into the ordinary air. The gases dissolved in his blood expand like the gas in a bottle of soda-water when the cork is drawn, and the bubbles interfere with the circulation of the blood in the finer blood-vessels (of especial importance being those of the brain and spinal cord), and the serious illness and the death of workmen has frequently resulted from this cause. Accordingly, the men who work in such "compressed atmospheres" are now made to pass slowly through a series of three chambers, in each of which the pressure is diminished and brought nearer to that of the normal atmosphere. By spending twenty minutes in each chamber successively, the workman is gradually brought to the pressure of the outer world, and his blood prevented from "effervescing." But what must be the condition of the gases in the blood of a whale which suddenly rises from 400 fathoms to the surface? The whale suddenly goes, not from a pressure of four times the normal ("four atmospheres," as it is called), but from eighty times the normal, to the normal pressure.

Whales, and also seals, are provided with remarkable special networks of blood-vessels in various parts of the body (called "retia mirabilia" by the old anatomists,) and also with a thick layer of fat under the skin, the "blubber" (some feet deep in a large whale), full of blood-vessels. It has been suggested that these networks of blood-vessels are related in some way both to the power of keeping long (forty minutes!) under water without breathing, and also to the freedom of these marine monsters from the deadly effects of rapid passage from great to little gas-pressure. But it is only a suggestion; no one has shown how the networks can act so as to effect these results, and I am quite unable to say how they do so. Another suggestion worth considering is that the whale completely empties the gas out of its lungs by muscular compression of the body-wall before diving, so that there is no gas left in the body to be acted on by the increased pressure resulting from its sinking into deep water. I am unable to deal with this puzzle myself, and I have not been able to find any naturalist or physiologist who can throw light on the matter.

The toothed whales are nearer to the ancestral primitive whales than are the whalebone whales. The latter are the more peculiar, and specially adapted with their huge heads and mouths (a third the length of the whole animal in the Greenland whale), and their palisades of 350 whalebone planks, some

12 ft. long, on each side of the mouth. I may mention in parenthesis that, whilst whalebone has been largely superseded by light steel in the making of umbrellas and corsets, its value remains, or rather increases, on account of its being the only material for making certain kinds of large brushes which are used in cleaning machinery. The whalebone whales have, when first born, very minute teeth hidden in their jaws; they disappear. Some of the toothed whales have teeth only in the lower jaw (the cachalot), others (the beaked whales, Ziphius, etc.) have only one pair or two pairs of teeth. These are tusk-like, and placed in the lower jaw. Others (the dolphins and porpoises) have very numerous peg-like teeth in each jaw. Some of them feed on fish, pursuing the shoals of fish in parties or "schools."

A truly terrible toothed whale is the large porpoise called the killer (known to zoologists as Orca gladiator). He is the wolf of the sea, far more active and formidable than any shark, about 10 ft. long, and strangely marked in black, white, and yellow. He has jaws bigger than those of the largest Mugger crocodile, and a tremendous array of fang-like teeth. These killers hunt the Right (or whalebone) whales in all parts of the world, in parties of three to twelve. They hang on to the lips of their enormous "quarry," and once they get a hold, in twenty minutes tear it into pieces. Often they satisfy themselves with tearing out and devouring the gigantic tongue of their victim, leaving the carcase untouched.

The narwhal and the white whale, or Beluga, which furnishes "porpoise-hide" for boots and laces, are both caught in northern seas, and form a closely allied pair, similar to one another in shape and colour (the one white, the other grey), and of moderate size, about 12 ft. long. They both feed on cuttle-fish and minute shrimps, but the Beluga has many teeth and the narwhal (with the exception of some rudimentary ones) only a single pair, and these in the front of the upper jaw. In the female narwhal their pair of teeth remain permanently concealed in the jaw bone, and so does the right side one of the male. But the left side tooth of the male grows to an enormous size, projecting horizontally in front of the narwhal to a length of seven or eight feet. It is a powerful weapon, and is formed of ivory spirally grooved on the surface. The narwhal was called "the unicorn fish" or "Monoceras" in ancient times, and its spirally marked tooth was confused with the horn of the terrestrial unicorn—the rhinoceros. Very rarely the right tooth of the male narwhal grows to full size side by side with the left tooth. A specimen showing this double-toothed condition is in the Natural History Museum. A most curious fact, quite unexplained as yet, is that the spiral grooving on both the teeth turns in the same direction; in both it is like a spiral staircase in mounting which (starting from the base implanted in the jaw) you continually turn to the right. Now, in all other animal structures which have a spiral growth and are paired—one belonging to the right side of the animal, the other to the left, as, for instance, the spirally marked horns of antelopes and the more loosely coiled horns of sheep and cattle—one of the pair forms a right-handed and the other a left-handed spiral. They are "complementary"; one is the reflection, as in a mirror, of the other. Why the narwhal's tooth does not conform to this rule is a mystery.

It is a remarkable fact that only a few whales and porpoises eat fish or the flesh of other whales. The large toothed-whales, including the cachalot or sperm whale, and also the Ziphius-like beaked whales, live upon cuttle-fish. And it seems that they know where to hunt for this special article of diet and how to find it in quantity (probably at great depths in the ocean), which naturalists do not. Many new kinds of cuttle-fish have been discovered by examining the contents of the stomach of captured whales. The sperm whale feeds on monster squid and poulp such as we rarely, if ever, see alive or washed up on the shore. The hide of these cuttle-fish-eating whales and porpoises is scratched and scarred by the hooks attached to the suckers on the arms of the great cuttle-fish, and a test of the genuine character of ambergris which forms as a concretion in the intestine of the sperm-whale is that it contains fragments of the horny beaks and hooks of the cuttle-fish digested by the whale. The food of the whalebone whales consists of minute crustacea and of the little floating molluscs known as Clio borealis, as big as the last joint of one's little finger, which float by millions in the Arctic Ocean. The whalebone whales, after letting their huge mouths fill with the sea-water in which these creatures are floating, squeeze it out through the strainer formed by the whalebone palisade on each side—by raising the tongue and floor of the mouth. The water passes out through the strainer, and the nourishing morsels remain.

Some fossil jaws and skulls of whales from miocene and older tertiary strata are known which tend to connect the toothed whales with those mammals not modified for marine life. But the approach in that direction does not go very far. The extinct whales called Squalodon have tusk-like front teeth and molars which have the outline of a leaf with a coarsely "serrated" edge. The bones of the face are also, in them, more like those of an ordinary mammal than is the case with modern toothed whales. The snout is not so long, and the bones which form it are a little more like those of a fox's snout than are those of the dolphin's "beak." But on the whole it is astonishing how little we know of fossil whales. We have yet to discover ancestral forms possessing small hind legs, but whale-like in other features. Some day a lucky "fossil-hunter" will come upon the remains of a series of whale-ancestors probably of Eocene age, and we shall know the steps by which a quadruped was changed into a cetacean—just as we have recently learned the history of the development of elephants. We know even less about the ancestry of bats and the steps by which they acquired their wings than we do about the history of whales. These discoveries await future generations of men when "cuttings" and "pits" and quarries shall have been made in the rest of the earth's surface to the same extent as they have been in Europe and in parts of the American continent.

MISCONCEPTIONS ABOUT SCIENCE

I submit, as the final chapter of this little volume of miscellaneous diversions, a few words intended to meet what has become a recurrent misrepresentation and absurdity for which the annual congress of the British Association for the Advancement of Science furnishes the opportunity. Glib writers in various journals regularly seize this occasion to pour forth their lamentations concerning the incapacity of "science" and the disappointment which they experience in finding that it does not do what it never professed to do. They deplore that those engaged in the making of that new knowledge of nature which we call "science" do not discover things which they never set out to discover or thought it possible to discover, although the glib gentlemen who write, with a false assumption of knowledge, pretend that these things are what the investigations of scientific inquirers are intended to ascertain. We read, at that season of the year, articles upon "What Scientists do not know" and "The Bankruptcy of Science," in which it is pretended that the purpose of science is to solve the mystery, or, as it has been called, the "riddle," of the universe, and it is pointed out, with something like malicious satisfaction, that, to judge by the proceedings of the congress of scientific investigators just concluded, we are no nearer a solution of that mystery than men were in the days of Aristotle: and it is added that false hopes have been raised, and that matters which were once considered settled have again passed into the melting-pot! This kind of lamentation is not only (if I may use an expressive term) "twaddle," but is injurious misrepresentation, dangerous to the public welfare. The actual attitude of the investigators and makers of new knowledge of nature is stated in a few words which I wrote ten years ago: "The whole order of nature, including living and lifeless matter—from man to gas—is a network of mechanism, the main features and many details of which have been made more or less obvious to the wondering intelligence of mankind by the labour and ingenuity of scientific investigators. But no sane man has ever pretended, since science became a definite body of doctrine, that we know or ever can hope to know or conceive of the possibility of knowing, whence this mechanism has come, why it is there, whither it is going, and what there may or may not be beyond and beside it which our senses are incapable of appreciating. These things are not 'explained' by science and never can be."

So much for those who reproach science with the non-fulfilment of their own unwarranted and perfectly gratuitous expectations.

When, however, having created in their readers' minds an unreasonable sense of failure and a mistrust of science, such writers go on to make use of the want of confidence thus produced, in order to throw doubt upon the real conquests of science—the new knowledge actually made and established by the investigators of the last century—it becomes necessary to say a little more. The public is told by these false witnesses that science has "dogmas," and that men of science are less satisfied than they were with the "dogmas" of the last century. Science has no dogmas; all its conclusions are open to revision by experiment and demonstration, and are continually so revised. But science takes no heed of empty assertion unaccompanied by evidence which can be weighed and measured. "Nullius in verba" is the motto of one of the most famous Societies for the promotion of the knowledge of nature—the Royal Society of London.

It is especially in the area of biology—the knowledge of living things—that the enemies of science make their most audacious attempts to discredit well-ascertained facts and conclusions. They tell their readers that those greater problems of the science (as they erroneously term them), such as the nature of variation among individuals, the laws of heredity, the nature of growth and reproduction, the peculiarities of sex, the characteristics of habits, instinct, and intelligence, and the meaning of life itself, have advanced very little beyond the standpoint of the first and greatest biologist, Aristotle. This statement is vague and indefinite; the conclusion which it suggests is absolutely untrue. Aristotle knew next to nothing about the mechanism of the processes in living things above cited. At the present day we know an enormous amount about it in detail. But when men of science are told that they do not know the "nature" of this and the "meaning" of that, they frankly admit that they do not know the real "nature" (for the expression is capable of endless variety of significance) of anything nor the real "meaning" not only of life, but of the existence of the universe, and they say, moreover, that they have no intention or expectation of knowing the ultimate "nature" or the ultimate "meaning" (in a philosophical sense) of any such things. These are not problems of science—and it is misleading and injurious to pretend that they are.

I recently read an essay in which the writer is good enough to say that, owing to the work of Darwin, the fact that the differences which we see between organisms have been reached by a gradual evolution, is not now disputed. That, at any rate, seems to be a solid achievement. But he went on to declare that when we inquire by what method this evolution was brought about biologists can return no answer. That appears to me to be a most extraordinary perversion of the truth. The reason why the gradual evolution of the various kinds of organisms is not now disputed is that Darwin showed the method by which that evolution can and must be brought about. So far from "returning no answer," Darwin and succeeding generations of biologists do return a very full answer to the question, "By what method has organic evolution been brought about?" Our misleading writer proceeds as follows: "The Darwinian theory of natural selection acting on minute differences is generally considered nowadays to be inadequate, but no alternative theory has taken its place." This is an entirely erroneous statement. Though Darwin held that natural selection acted most widely and largely on minute differences, he did not suppose that its operation was confined to them, and he considered and gave importance to a number of other characteristics of organisms which have an important part in the process of organic evolution. The assertion that the theory of natural selection as left by Darwin "is now generally held to be inadequate" is fallacious. Darwin's conclusions on this matter are generally held to be essentially true. It is obvious that his argument is capable of further elaboration and development by additional knowledge, and always was regarded as being so by its author and by every other competent person. But that is a very different thing from holding Darwin's theory of natural selection to be "inadequate." It is adequate, because it furnishes the foundation on which we build, and it is so solid, complete and far-reaching that what has been added since Darwin's death is very small by comparison with his original structure.

Lastly, we are told by the anonymous writer already quoted that at the present time discussion is chiefly concentrated on the question as to whether life is dependent only on the physical and chemical properties of the living substance, protoplasm, or whether there is at work an independent vital principle which sharply separates living from non-living matter! And the obvious and common-place conclusion is announced that "the ultimate problems of biology are as inscrutable as of old." All ultimate problems are, I admit, inscrutable. It is, on the other hand, the business, and has been the glory and triumph, of science, to examine and solve problems which are scrutable! It is certainly not the case that, at the present time, discussion is concentrated on the question of the existence of a vital principle. There is absolutely no discussion in progress on the subject. No one even knows or attempts to state what is meant by "a vital principle." It is a phrase which belongs to "the dead past," when men of science had not discovered that you get no nearer to understanding a difficult subject by inventing a name to cover your ignorance. Thirty-five years ago the word "vitality" was used as some few philosophising writers are now using the term "vital principle." Huxley at that time attacked the views of Dr. Lionel Beale, who called in the aid of a mystical "principle," which he named "vitality," in order to "account for" some of the remarkable properties of protoplasm. As Huxley pointed out, this supposed principle "accounted for" nothing, since it was merely a name for the phenomena for which it was supposed to account. Huxley pointed out that many chemical compounds have remarkable properties—as assuredly have the chemical compounds which are present in protoplasm—but men of science have not found it to help them in investigating the mechanism of those properties to ascribe them to mystical intangible "principles" differing from the agencies at work in other less exceptional substances.

Thus, for instance, water, though a very common and abundant chemical compound formed by the union of two chemical elements, hydrogen and oxygen, which, at the temperature and pressure of the earth's surface, are gaseous, offers many strange properties to our consideration not shared by other compounds of gaseous elements. For instance, hydrogen, when it combines with gaseous elements other than oxygen, does not form a compound which is liquid at the temperature and pressure of the earth's surface. Its combinations with nitrogen, with chlorine, with fluorine, and even some with the solid element carbon, are under those conditions gaseous. What a special character, therefore, has water! Moreover, water, though a liquid, yet behaves in a most peculiar way when either cooled below ordinary temperatures or heated above them. It becomes solid when cooled, but expands at the same time, so that it is less dense when solid than when liquid—a most unusual proceeding! And when heated it is converted into vapour, but with a loss or "making latent" of heat, which, like its behaviour when solidifying, indicates that water is endowed with a very peculiar structure or mechanism in the putting together of its molecules. We might call these combined peculiarities of water "aquosity," and as we certainly cannot say why water should possess the lot of them, whilst other compounds of either hydrogen or of oxygen, or, in fact, of any other elements, do not possess this combination, we might say that their presence is due to "the aqueous principle," or "aquosity," which enters into water when it is formed, but does not exist in other natural bodies, and, indeed, "sharply separates aqueous from non-aqueous matter."

Happily, though such a view would have been considered high philosophy 200 years ago, no one is deluded at the present day into the belief that by calling the remarkable properties of water "aquosity" you have added anything to our knowledge of them. Yet those who invoke "a vital principle" or "vitality" in connection with protoplasm should, if they were consistent, apply their method to the mystery of water. Let us see how it would run. Though we may (the "vitalists" or "aquosists" would say) experiment with water, determine exactly the temperature and pressure at which these remarkable phenomena are exhibited, though we may determine its surface tension and its crystalline form, and even though we may weigh exactly the proportion of hydrogen to oxygen in its composition, yet when we look at a drop of water, there it is, a wonder of wonders, endowed with "aquosity," the ultimate nature of which is as inscrutable now as it was to Aristotle! It is perfectly true (we concede to the "aquosists") that the properties of water are not accounted for by science; that is to say that, though we can imagine the molecular and atomic mechanism necessary for their exhibition, we cannot offer any suggestion as to how it is that that particular mechanism is present in the chemical compound which the chemist denotes as H2O, and is not present in other compounds, still less can we say "why" these remarkable properties are present—that is to say, for what purpose, although we know that if they were not present the whole history and economy of our globe would be utterly different from what it is. Nevertheless, in spite of their ignorance about the real nature of water, men of science do not invent an "aqueous principle" or "aquosity" with the notion of "explaining" water. And I have yet to hear of any duly trained and qualified biologist who is prepared at the present moment to maintain the existence of a "vital principle," or of a force to be called "vitality," supposed to be something different in character and quality from the recognised physical forces, and having its existence alongside, yet apart from, the manifestations of those forces.

Lord Justice Fletcher Moulton recently said: "The advance in science takes the workers in science more and more beyond the ken of the ordinary public, and their work grows to be a little understood and much misunderstood; and I have felt that, as in many other cases, the need would come for interpreters between those who are carrying on scientific research and the public, in order to explain and justify their work." Probably everyone will agree with the Lord Justice: but what are we to say of those responsible owners of great journals who not only abstain from providing such interpretation but allow anonymous and incompetent writers to mislead the public? Is the literary critic of a prosperous journal employed to write the City article?

There has been a repetition this year (1912) of the usual misrepresentation on the occasion of the meeting of the British Association. The President, Professor Schäfer, had let it be known that his address would be concerned with the chemistry of living processes, the gradual passage of chemical combinations into the condition which we call "living," and the possibility of bringing about this passage in the chemical laboratory without the use of materials already elaborated by previously existing "living" material. The announcement was immediately made in some "newspapers" that "startling revelations" were to be made by the President, that he was "to throw a bomb-shell" into the camp, etc. He did nothing of the kind. He gave an admirable and clear statement of the progress during recent years towards the realisation of the construction in the laboratory by chemical methods of the complex chemical combination which exhibits those "activities"— essentially movements, unions, disruptions and re-unions of extremely minute particles—which we call "living." The conclusion that such a gradual building up has taken place in past ages of the history of our earth was formulated more than forty years ago by Spencer, Tyndall, Huxley, Haeckel, and others, and has not been seriously attacked in the interval, but, on the contrary, generally accepted as a legitimate inference from the facts ascertained and the theory of the evolution or gradual development of what we call the material universe.

Professor Schäfer expressed the opinion, anticipated and shared by many other investigators, that the progress of chemical experiment renders it probable that further steps, culminating in the successful construction of "living" matter in the laboratory, are not beset by any insurmountable obstacles and will sooner or later be accomplished. There was no "bomb-shell" in this statement, and no excitement as its result among scientific workers nor amongst those who do not neglect to study the writings of the "interpreters" desired by Lord Justice Moulton. There are still some such interpreters carrying on the work of Huxley and of Tyndall, those great interpreters whose writings should be studied and treasured as classics.

The most interesting result of the attempt to treat the discussions at Dundee as a newspaper "sensation," comparable to the reports relating to motor-car bandits or the pronouncements of political factions, has been its complete failure. Serious thinkers of all schools seem to have adjusted themselves to the more modern way of regarding natural processes even when these relate to matters of such age-long interest to mankind as the inception of "living" organisms and of conscious humanity itself. There are fewer now than there were forty years ago who insist on the older barbaric "explanations" of these marvels. Few indeed venture to assert the existence of "spirits"—ghostly essences of various grades and capacities which enter the bodies of living things and escape from them like so much gas when they die.[10] The vegetable soul, the animal soul and the human soul are no longer imagined and described to us as definite "things" supposed to "explain" the complex processes which go on respectively in plants, animals and men.

Seventy years ago the facts which were known as to that changing state of material substances which we describe by the words "hot" and "cold," were held to be "explained" by the existence of a ghostly thing called "caloric," which was believed to enter various bodies and make them hot and then to escape from them and so make them cold. Primitive man multiplied such ways of explaining each and every process going on in the world around him and in himself. Mere words or names lost their first simple signification and acquired permanent association with imaginary spirits, demons, and haunting intangible ghosts, by reference to which our ancestors in their earliest "reasoning" explained to their own satisfaction the strange and sudden events fraught to them with the daily experience of pain or pleasure. The whole world was held by them to be "bewitched," and it was only by slow and painful steps that some knowledge of the persistent order of Nature was obtained, whilst the phantastic imagery which had served in its place, bit by bit disappeared. "Caloric" was a late lingerer, and was only got rid of when what had been so called was shown to be a vibration of particles—a mode or kind of motion—a "state," and not a mysterious fluid existing as a thing in itself.

Just as "caloric" no longer serves and is no longer possible as the supposed "explanation" of the behaviour of bodies in the hot or the cold state, so we no longer require the supposition of "spirits" of one kind or another as "explanations" of the living state of those products of our mother earth which are called plants, animals and men. In neither case do such "spirits" really "explain" the state in question; they are only names for the activity which it was imagined that they served to explain. These states or affections of matter remain as wonderful and important to us as they were before. But by giving up the prehistoric notions about them which have been handed on until the present day we can think of them in a more satisfactory way—a way which avoids the multiplication of unnecessary imaginary agencies and the conception of an intermittent and hesitating Creative Power, and substitutes for it the operation of continuous orderly and preordained forces.

It is true that we can neither ascertain nor imagine either the beginning or the end of the orderly process which we discover in operation to-day. We can trace it back by well-established inference into a remote past, but a beginning of it is not within the possibilities of human thought. We can, with reasonable probability of being correct, foretell the changes and developments which time will bring in many combinations and dispositions which are the manifestations of that process at this moment of time, but we cannot even think of a cessation of that process.

Should we ask, "Why does this process exist?" there is no answer. Nature does not reply; an awful silence meets our inquiry. The reproach is often urged against science—the knowledge of the order of nature—that it does not tell us "why we are here." Man inevitably desires to know why he is here; but "science," as that word is now understood, does not profess or even seek to answer that question, although the false hope has been raised in ignorant minds, sometimes by knavery, sometimes by honest delusion, that it could do so. By knowledge of nature mankind can escape much suffering and gain the highest happiness, but that is all that we can hope for from it. We shall never satisfy our curiosity; we shall never know in the same way as we know the order of nature, why—to what end, for what purpose—that order and not another order exists.

It is very generally supposed that it is the business and profession of science "to explain" things—that is to say, to show how this or that must and does come about in consequence of the operation of the great general properties of matter, known as the "laws" of chemistry and physics. This is true enough, but it is equally the work of science to assert that of many things for which mankind demands "an explanation," there is no explanation. It is further the work and the service of science to destroy and to remove from men's minds the baseless and pretended "explanations" which are no explanations but causes of error, blindness, and suffering.

Science, the destroyer of "explanations," is the purifier of the human mind, its cleanser from the crippling infection of prehistoric error and from domination by the terrifying nightmares of our half-animal ancestry.

Finally, in reference to the very ancient attempt to "explain" life and consciousness by the assertion that they are due to "spirits" which enter the bodies of animals and men, I must caution the reader against supposing that—for those who do not accept the belief that such spirits exist—the gravity and mystery of the manifestations of life and consciousness are in any way lessened. Those who reject the belief in "spirits" do not in consequence reject the ethical and moral doctrines which have too long been rendered "suspect" by the shadow cast over them by ancient superstition. The disappearance of that shadow will reveal friends where enemies were supposed to be entrenched.

At the meeting of the British Association in 1879 I delivered an address on "Degeneration: a Chapter in Darwinism." In the printed version of that address, published in the same year, there are some statements bearing on the matter above discussed which I reproduce here, since I can still make them with conviction.

"Assuredly it cannot lower our conception of man's dignity if we have to regard him as 'the flower of all the ages' bursting from the great stream of life which has flowed on through countless epochs with one increasing purpose, rather than as an isolated miraculous being, put together abnormally from elemental clay, and cut off by such portentous origin from his fellow animals and from that gracious nature to whom he yearns with filial instinct, knowing her, in spite of fables, to be his dear mother."

"A certain number of thoughtful persons admit the development of man's body by natural processes from ape-like ancestry, but believe in the non-natural intervention of a Creator at a certain definite stage in that development, in order to introduce into the animal which was at that moment a man-like ape, something called 'a conscious soul' in virtue of which he became an ape-like man."

"No one ventures to deny, at the present day, that every human being grows from the egg in utero, just as a dog or a monkey does; the facts are before us and can be scrutinised in detail. We may ask of those who refuse to admit the gradual and natural development of man's consciousness in the ancestral series, passing from ape-like forms into indubitable man, 'How do you propose to divide the series presented by every individual man in his growth from the egg? At what particular phase in the embryonic series is the soul with its consciousness implanted? Is it in the egg? in the fœtus of this month or that? in the new-born infant? or at five years of age?' This, it is notorious, is a point upon which churches have never been able to agree; and it is equally notorious that the unbroken series exists—that the egg becomes the fœtus, the fœtus the child, and the child the man. On the other hand we have the historical series—the series, the existence of which is inferred by Darwin and his adherents. This is a series leading from simple egg-like organisms to ape-like creatures, and from these to man. Will those who cannot answer our previous inquiries undertake to assert dogmatically in the present case at what point in the historical series there is a break or division? At what step are we to be asked to suppose that the order of nature was stopped, and a non-natural soul introduced?... The theologian is content in the case of individual development of the egg to admit the fact of individual evolution, and to make assumptions which lie altogether outside the region of scientific inquiry. So, too, it would seem only reasonable that he should deal with the historical series, and frankly accept the natural evolution of man from lower animals, declaring dogmatically, if he so please, but not as an inference of the same order as are the inferences of science, that something called the soul arrived at any point in the series which he may think suitable. At the same time, it would appear to be sufficient even for the purposes of the theologian, to hold that whatever the two above-mentioned series of living thing contain or imply, they do so as the result of a natural and uniform process of development, that there has been one 'miracle' once and for all time....

"The difficulties which the theologian has to meet when he is called upon to give some account of the origin and nature of the soul certainly cannot be said to have been increased by the establishment of the Darwinian theory. For from the earliest days of the Church, ingenious speculation has been lavished on the subject.

"St. Augustine says (I give a translation of the Latin original): 'With regard to the four following opinions concerning the soul—viz. (1) whether souls are handed on from parent to child by propagation; or (2) are suddenly created in individuals at birth; or (3) existing already elsewhere are divinely sent into the bodies of the new-born; or (4) slip into them of their own motion—it is undesirable for anyone to make a rash pronouncement, since up to the present time the question has never been discussed and decided by catholic writers of holy books on account of its obscurity and perplexity—or, if it has been dealt with, no such treatises have hitherto come into my hands.'"

There must be many who will be glad to shake off the illusion of explanation which is no explanation, and to escape from the futile discussion of the possible behaviour of spirits and ghosts born in the dreams of primæval savages. They will gladly accept the conclusion that the marvellous qualities and activities of living things and that inscrutable wonder, the mind of man, are outcomes of the orderly process of Nature no less than are the miracles which we call a buttercup, a rock crystal, a glacier, the noon-day sun! We can trace, by observation and inference, the orderly growth and development of these things from simpler things; we can discover continuity and common properties determining their diverse existence. But we find no explanation of them; we cannot account for the properties of matter which determine them, nor for the existence of anything—whether it be a drop of water, or human thought and consciousness. There are no special and exceptional "incomprehensibles" requiring us to assume that special "principles" or "spirits" are concerned with them whilst the rest are to be accounted for and explained in a more general way. Wherever we push our inquiries we come equally and inevitably, as did primæval man, to that of which there is no explanation— the perpetual miracle, the miracle of the nature of things, of existence itself. The man of science bows his head in the presence of this all-pervading mystery. He is called arrogant by those who arrogate to themselves the right to "explain" things and to deal in vital spirits and metaphysical nostrums for that purpose. From time to time they fill with their proclamations the great silence which he has learnt to accept with reverence and humility. As the years roll on their hollow phrases are less frequent, and acquire the pathetic interest which belongs to all such decaying remnants of the thought and effort of the childhood of man.

It seems still to be necessary to insist that it is not reasonable to assume as an indisputable fact that man can arrive at an "explanation" of existence and the nature of things. This assumption has been made in the past, and, by a well-known trick of advocacy, it has been argued that since science fails to "explain" these things, the old prehistoric fancies as to spirits—even though they "explain" nothing and have themselves to be "explained"—hold the field and must be accepted as true. There is an alternative, and that is to admit our ignorance. No man has ever seen or knows what is on the other side of the moon, that which does not face our earth. There are few amongst us who, in this admitted and complete state of ignorance, would persist in declaring that we must accept as true the suppositions of ancient races of men as to the existence there of men-like creatures, or would be deluded by the argument that since we do not know what is there the suppositions in question must be accepted as true. We cannot, as a matter of observation, assert that these supposed beings are not there, but we can find no reason to make it appear even probable, nor any means of proving by experiment, that they are. We refuse to entertain such suppositions.

[10] This subject is discussed and some account of the chemical nature of protoplasm given in my book, "Science from an Easy Chair" (Methuen, 1910), which consists of a first series of papers similar to those which are collected in the present volume as a "Second Series." The chapters in the earlier volume to which I wish to direct the reader's attention are those entitled "The Universal Structure of Living Things," "Protoplasm, Life and Death," "Chemistry and Protoplasm," "The Simplest Living Things."

www.ingramcontent.com/pod-product-compliance
Lightning Source LLC
Chambersburg PA
CBHW051902170526
45168CB00001B/215